W0080706

Textbook of
Cosmetics

Textbook of
Cosmetics

M. Vimaladevi

B.Sc., M.Pharm., Ph.D.

Emeritus Professor of Pharmaceutical Technology,
Andhra University, Visakhapatnam (A.P.)

Honorary Chairperson,
Auro Pharma, Pondicherry

CBSPD

CBS Publishers & Distributors Pvt Ltd

New Delhi • Bengaluru • Chennai • Kochi • Kolkata • Lucknow • Mumbai
Hyderabad • Jharkhand • Nagpur • Patna • Pune • Uttarakhand

> The work is offered to
> **THE MOTHER & SRI AUROBINDO**

Disclaimer

Science and technology are constantly changing fields. New research and experience broaden the scope of information and knowledge. The authors have tried their best in giving information available to them while preparing the material for this book. Although all efforts have been made to ensure optimum accuracy of the material, yet it is quite possible some errors might have been left uncorrected. The publisher, the printer and the authors will not be held responsible for any inadvertent errors, omissions or inaccuracies.

> **Textbook of**
> **Cosmetics**

ISBN-13: 978-81-239-1103-8

Copyright © Author and Publisher

First Edition: 2005
   Reprint: 2006, 2007, 2009, 2010, 2011, 2012, 2015, 2017, 2019, 2023, 2025

All rights reserved. No part of this book may be reproduced or transmitted, in any form or by any means, electronic, mechanical, photocopying, recording or any information storage and retrieval system without the permission, in writing, from the author and the publisher.

Published by **Satish Kumar Jain** and produced by **Varun Jain** for

CBS Publishers & Distributors Pvt Ltd
4819/XI Prahlad Street, 24 Ansari Road, Daryaganj, New Delhi 110 002, India.
Ph: 011-23266838, 23289259   Website: www.cbspd.com
                                 e-mail: delhi@cbspd.com

Corporate Office: 204 FIE, Industrial Area, Patparganj, Delhi 110 092
Ph: 011-4934 4934   Fax: 011-4934 4935
                           e-mail: publishing@cbspd.com; publicity@cbspd.com

Branches

- **Bengaluru:** Seema House 2975, 17th Cross, KR Road, Banasankari 2nd Stage, Bengaluru 560 070, Karnataka, India
  Ph: +91-80-26771678/79   Fax: +91-80-26771680   e-mail: bangalore@cbspd.com
- **Chennai:** 7, Subbaraya Street, Shenoy Nagar, Chennai 600 030, Tamil Nadu, India
  Ph: +91-44-26680620, 26681266   Fax: +91-44-42032115   e-mail: chennai@cbspd.com
- **Kochi:** 42/1325, 1326, Power House Road, Opp KSEB, Power House, Ernakulam Kochi 682 018, Kerala, India
  Ph: +91-484-4059061-65,67   Fax: +91-484-4059065   e-mail: kochi@cbspd.com
- **Kolkata:** 147, Hind Ceramics Compound, 1st Floor, Nilgunj Road, Belghoria, Kolkata-700056, West Bengal, India
  Ph: +033-25633055, 033-25633056   e-mail: kolkata@cbspd.com
- **Lucknow:** Basement, Khushnuma Complex, 7 Meerabai Marg (Behind Jawahar Bhawan), Lucknow-226001, UP India
  Ph: +0522-4000032   e-mail: tiwari.lucknow@cbspd.com
- **Mumbai:** PWD Shed, Gala no 25/26, Ramchandra Bhatt Marg, Next to JJ Hospital Gate no. 2, Opp. Union Bank of India, Noorbaug, Mumbai-400009, Maharashtra, India
  Ph: 022-66661880/89   e-mail: mumbai@cbspd.com

Representatives

• Hyderabad	0-9885175004	• Jharkhand	0-9811541605	• Nagpur	0-8692091830
• Patna	0-9334159340	• Pune	0-9664372571	• Uttarakhand	0-9716462459

Printed at Neekunj Print Process, Haryana, India

Foreword

I⊤ GIVES ME great pleasure to write this Foreword for this path-breaking book on cosmetics, the framework of which the Author had discussed with me two years back when she visited me at AICTE. I am happy to notice that the book is much more comprehensive than the framework she had initially discussed with me. The book covers the entire gamut of cosmetics, including their formulations, in such a way that both the students as well as the actual formulators will find it equally useful. The theory and practical aspects of cosmetics technology have been beautifully balanced in treatment in the book.

It is my earnest hope that the students will greatly benefit from this book during their studies as well as in their professional carriers.

I compliment the Author for her leadership and commitment in producing this book.

Prof R.S.Gaud
Senior Advisor
AICTE, New Delhi

Preface

THIS VOLUME IS A compilation of cosmetic products of body and beauty care, products intended to be of theoretical or practical help to students at graduate level of pharmacy.

Although written primarily with students in mind, the book will be also useful for the cosmetic industry, cosmetologists, beauticians, dermatologists, physicians and non-technical persons interested in knowing about cosmetics as it discusses in detail manufacturing processes of various cosmetics.

On the whole, the aim of this book is to give a comprehensive idea on cosmetics of the east and the west, their method of manufacture, packaging and application.

In view of the vast progress cosmetic technology has made in the recent past, there is a need for graduates to go deeper into this part of pharmaceutical technology in an unprecedented manner, specially with the revised technical syllabus including a separate paper on cosmetics which was earlier covered under pharmaceutical technology.

I regret the number of textual errors that have crept into the first print and have taken care to correct them in this print. I sincerely hope that this book serves its purpose. Suggestions from readers are welcome.

Acknowledgement

I ACKNOWLEDGE WITH GREAT fondness the encouragement and help my late husband Dr. R.V. Krishna Rao and children have provided me at all stages of preparation of this volume and thereafter at the time of its production. I greatly admire and cherish their support without which my ambition of writing this book would have remained a distant dream.

I am indebted to my typist Mrs. Jayashree, Mr Sridhar, Vijayaletchumi and Backialakshmi for maintaining their patience with me in accommodating all those changes which were called by me from time to time. I thank them for their understanding and perseverance.

Special thanks to my publishers who gave my work the form of a book thus helping me fulfil my aim.

Contents

Introduction

COSMETICS ARE IN use since thousands of years. The practice continues today, because of growing importance in personal body and beauty care in many parts of the world. A growing interest among consumers in cosmetics has created the need for greater precision and scientific back up in the development and manufacturing processes of products. This has urged the pharmacists to identify pharmaceutical technology as one of the relevant subjects for the future development of cosmetics.

Cultural beliefs and traditional practices have played a substantial role in continued application of cosmetics, in the West as well as in the East. This made the scientists carry out research in cosmetics, and as a result more and more products are being developed and marketed. Body and beauty care products are likely to surpass the consumption of drugs in future. A large segment of the world population is showing greater inclination towards natural cosmetics which seems to be the future hope.

Throughout the ages cosmetics have played a major role as tools of beautification for men and women alike. Medicinal and cosmetic arts were quite advanced in many civilizations and India was no exception. Infact, Sushruta a pioneer in surgery, (6th century A.D.) is also remembered for his suggestions on maintaining a healthy body, diet, hygiene and exercise.

India was the leader in medicinal and cosmetic arts during the Gupta period (3rd to 5th century A.D.). Cosmetics and hygiene were advanced. The daily toiletries of ladies included many types of creams, oils and pastes like allagech (Eagle Wood) costus, frankincense, myrrh, camphor, saffron and sandalwood—which are some of the oldest known aromatic substances from India.

Make-up of many kinds including hair dyes, were used by men and women alike. Attars and floral waters for bath and a list of cosmetics exclusively for men also existed.

Betel leaves were used to darken lips and teeth; vermilion and other colours with waxes were used for facial designations. Almond paste for the entire body, perfumes and aromatics were used in all forms for religious rites and on social occasions such as weddings, birthdays etc.

Apart from the Indian civilization there were many others that developed the art of cosmetics-Egyptian, Persian, Greek, Chinese and Japanese. Travellers to and from India, China, Japan and Tibet spread much knowledge of cosmetics worldwide.

The Persians and the Arabs had improved methods of treating condition of the skin and hair. They had even enriched their materia medica with plants brought from Asia. Perfumes were introduced in Rome first by barbers as solid unguents, oily liquids and powders. Some perfumeries were even selling cosmetic appliances if we may call them so, useful in the manufacture of cosmetics.

Since the 16th century, cosmetics were separated from medicine and from the 17th century, cosmetic technology started developing on its own. Cosmetics were defined as:

- Preparations and not devices.
- Cleaning agents, like shaving creams, shampoos, cleansing creams and lotions etc.
- Products that can include drugs, e.g. antiperspirant creams and
- Products to be used on the human body.

Today, a plethora of cosmetic applications, for every part of the body is available in the world market. The evolution of cosmetics proves that there is a clear distinction between the earlier drama artist's make-up material and popular cosmetics for general usage.

Face and
Body Powders

1

Of ALL THE COSMETICS used, face powder easily sits on the top of the list. It is equally popular among men as well as women. It was a common practice among women in the olden days to use a little cornstarch or flour on their faces to cover shine and for sometime just plain white powder was very popular. Technological developments in the manufacture of face powder lead to incorporation of colour and perfume. The modern day face powder is a product that adds very materially to personal beauty. People who are conscious about their looks apply it several times a day. Though the applications are small, the total consumption is high.

Due to misconception and also due to competitive and negative type of advertising, the use of face powder came under some criticism. Bismuth, lead or mercury contained in it were believed to be harmful. Furthermore, it was thought to be the cause of pimples and blackheads, that it gets into pores of the skin and clogs them, that orris root in particular which it contained, produced an allergy like hay fever, or asthma and the use of starch was harmful to the skin.

Modern face powders are manufactured by reputed companies using purified ingredients, approved colours and non-irritating perfumes. Inspite of daily applications none of these harmful effects have been suffered.

A face powder manufacturer's priority concerns the requirements of women. Fragrance is the first characteristic, because the first thing a woman does is to smell the powder while examining it. The nature of the fragrance should be such that it blends well with other perfumes and is not too predominant. It should also have lingering qualities both in the sealed or opened package and upon the skin. Next, the shade is very important as a large percentage of women buy face powder exclusively because of its shade. A large variety of shades are available with the approval of federal agencies like FDC (Food Drug & Cosmetic) or D&C (Drug & Cosmetic). As a rule, the manufacturer selects water insoluble colours, pigment and lakes for best results.

Apart from colour and odour a face powder must possess, slip, adhesion, fine particle size and covering power for consumer acceptance. Furthermore, it should be soft and fine to feel and have proper bulking powder. The various grades of face powder available are light, medium and heavy. Light and medium are the most popular covering powders.

The terms 'light', 'medium', and 'heavy' used for covering powder really have nothing to do with the density of the powder. These terms refer to the amount of powder needed to cover the same area of skin.

The types of skin which the face powder must cover are dry, normal, moderately oily, and very oily. Dry skin secretes very little moisture and no oil. Therefore it requires a powder with light covering power. Normal and moderately oily skin being more shiny due to secretion of moisture and oil requires a powder with more covering power. Very oily skins require a powder of heavy covering power due to high shine.

There are many women with generally dry skin and yet suffer from oily foreheads, oily noses or oily chins. In such cases, different powders of different covering powers are simultaneously used.

In addition to the covering powder the manufacturer must also provide suitable perfume, shades and attractive packaging.

Intelligent marketing strategy may recommend women to buy two colours—one for the day and the other lighter one for the night to counter the effects of artificial lighting.

PROPERTIES OF FACE POWDER

Let us try to understand the different properties of a face powder.

Covering Powder

One of the important functions of a face powder is to conceal the shine and minor skin imperfections. For this the primary requisite is covering power. Most raw materials possessing this particular property have it excessively and are somewhat lumpy and tend to drag when applied to the skin. It then becomes necessary to employ another material which serves the twin purpose of acting as a vehicle or diluent and at the same time facilitates easy application to the skin. Talc fits in suitably.

Next, adhesiveness of the skin must be sufficiently taken care of so as to enable the powder to remain on the skin for a considerable length of time. With magnesium, calcium and zinc stearates and sometimes with the addition of oil, this is achieved. Finally it must have an ingredient that acts as a binder for the colour and perfume and which facilitates distribution of both of them during manufacturing operation. Magnesium carbonate and precipitated chalk are used for this purpose.

A face powder must therefore possess the following properties.

(a) Covering power
(b) Slip
(c) Adhesiveness
(d) Absorption
(e) Colour and Perfume
(f) Should not give peach or mat like appearance after application
(g) Should not run off the face on motor riding
(h) Should not make the consumer apply many times on the face

The various weights of the face powder are proportional to the covering agent present in the formula. Basically a face powder can be made of any pure white, chemically inert powder that is sufficiently fine and exhibits one or more of the above properties.

The materials commonly used are given below.

Covering Power

Titanium dioxide, zinc oxide, zinc sulfide, lithopone, kaolin (colloidal), calcium sulfate, magnesium oxide, and starch.

Slip

Talc, starch, and metallic soaps.

Adhesiveness

Calcium, magnesium and zinc stearate, starch and colloidal clay (purified kaolin).

Absorption

Precipitated chalk, magnesium carbonate, starch, and purified kaolin.

A bulk of the commercial face powder constitutes the materials listed above. The principal materials are zinc oxide, titanium dioxide, talc, kaolin, the stearates, starch and chalk.

Rice starch is the ideal material possessing all the properties of face powder. However, it has not been favored in the USA due to its clinging properties which emphasizes light hairs on the lips and cheeks and its tendency to absorb moisture and smell although it does impart a much sought after smooth "peachy" effect. Nevertheless many leading French powders contain rice starch. Rice starch because of its drawback can be substituted by precipitated chalk. It imparts the same peachy texture and is not harmful. But ordinary grades do not spread or adhere evenly. Certain grades of processed chalk with all the required properties for a face powder can be used.

Chemists were apprehensive about using Barium Sulphate in view of the known toxicity of the other related barium salts.

Bismuth compounds are used in Europe although there is a disadvantage—they darken when exposed to light. Materials most widely used in Europe are zinc oxide, purified kaolin, talc, the metallic stearates—zinc stearate in particular—precipitated chalk and magnesium carbonate. Titanium dioxide is a more recent addition and is gaining popularity. It is being preferred to zinc oxide because it has better covering power than the latter, besides being extremely fine, white and chemically inert. Any insoluble, inert, white powder (about 200 mesh) having the desired properties can be used in the formulation.

With a better understanding of the densities of the face powder and a thorough knowledge of the ingredients that constitute it we move on to the development of formulas. It will be observed that the constant factors in each formula are covering agents, adhesive agents and absorbent agents. The variable factor is talc, that varies from formula to formula depending upon the amount of colour and perfume introduced.

Formulas

Light Face Powder with Zinc Oxide for Popular Shades

Materials	White	Peach	Rachel	Natural	Flesh	Ochre	Suntan
(a) Zinc oxide	18.00	18.00	18.00	18.00	18.00	18.00	18.00
(b) Talc	69.00	67.52	68.00	68.06	68.64	64.84	58.69
(c) Zinc stearate	6.00	6.00	6.00	6.00	6.00	6.00	6.00
(d) Precipitated Chalk	6.00	6.00	6.00	6.00	6.00	6.00	6.00
(e) Perfume	1.00	1.00	1.00	1.00	1.00	1.00	1.00
(f) Ochre	—	1.44	1.00	0.88	0.33	4.09	8.89
Brilliant pink lake	—	0.04	0.00	0.06	0.03	0.07	0.35
Burnt Sienna	—	0.00	0.00	0.00	0.00	0.00	1.07
	100.00	100.00	100.00	100.00	100.00	100.00	100.00

Medium Face Powder with Purified Kaolin with Popular Shades

Materials	White	Peach	Rachel	Natural	Flesh	Ochre	Suntan
(a) Titanium dioxide	4.00	4.00	4.00	4.00	4.00	4.00	4.00
(b) Purified Kaolin	20.00	20.00	20.00	20.00	20.00	20.00	20.00
(c) Talc	66.00	64.52	65.00	65.06	65.64	61.84	55.69
(d) Magnesium stearate	3.00	3.00	3.00	3.00	3.00	3.00	3.00
(e) Magnesium carbonate	6.00	6.00	6.00	6.00	6.00	6.00	6.00
(f) Perfume	1.00	1.00	1.00	1.00	1.00	1.00	1.00
(g) Ochre	—	1.44	1.00	0.88	0.33	4.09	8.89
Brilliant pink lake	—	0.44	0.00	0.06	0.03	0.07	0.35
Burnt Sienna	—	0.00	0.00	0.00	0.00	0.00	1.07
	100.00	100.04	100.00	100.00	100.00	100.00	100.00

Medium Face Powder with Titanium Dioxide for Popular Shades

Materials	White	Peach	Rachel	Natural	Flesh	Ochre	Suntan
(a) Titanium dioxide	5.21	5.21	5.21	5.21	5.21	5.21	5.21
(b) Talc	82.79	81.31	81.79	81.85	82.43	78.63	72.48
(c) Zinc stearate	5.00	5.00	5.00	5.00	5.00	5.00	5.00
(d) Precipitated Chalk	6.00	6.00	6.00	6.00	6.00	6.00	6.00
(e) Perfume	1.00	1.00	1.00	1.00	1.00	1.00	1.00
(f) Ochre	—	1.44	1.00	0.88	0.33	4.09	8.89
Brilliant pink Lake	—	0.44	0.00	0.06	0.03	0.07	0.35
Burnt Sienna	—	0.00	0.00	0.00	0.00	0.00	1.07
	100.00	100.04	100.00	100.00	100.00	100.00	100.00

Heavy Face Powder with Zinc Oxide and Titanium Dioxide for Popular Shades

Materials	White	Peach	Rachel	Natural	Flesh	Ochre	Suntan
(a) Titanium dioxide	3.00	3.00	3.00	3.00	3.00	3.00	3.00
(b) Zinc oxide	20.00	20.00	20.00	20.00	20.00	20.00	20.00
(c) Talc	66.00	64.52	65.00	65.06	65.64	61.84	55.69
(d) Zinc stearate	4.00	4.00	4.00	4.00	4.00	4.00	4.00
(e) Precipitated Chalk	6.00	6.00	6.00	6.00	6.00	6.00	6.00
(f) Perfume	1.00	1.00	1.00	1.00	1.00	1.00	1.00
(g) Ochre	—	1.44	1.00	0.88	0.33	4.09	8.89
Brilliant pink Lake	—	0.44	0.00	0.06	0.03	0.07	0.35
Burnt Sienna	—	0.00	0.00	0.00	0.00	0.00	1.07
	100.00	100.04	100.00	100.00	100.00	100.00	100.00

It may be noted that colours purchased in fairly large quantities ensure uniformity over a considerable period of time. Further,

1. All ingredients, with the exception of talc, are constant for all shades.
2. Chalk is a constant factor for all shades and weights.
3. The weight of the powder has no influence over the colour necessary to produce a given shade.
4. The light powder contains more zinc stearate than medium and heavy powders because less adhesiveness is required as the skin gets more and more oily.
5. Formulas employing zinc oxide as covering agent have talc content as constant.

This condition is desirable because it provides the heavy powder as much slip as a light powder. Many commercial powders have increased the zinc oxide content and decreased the talc content for greater covering power only to lose slip.

This problem can be rectified by increasing the covering power by the addition of a small quantity of titanium dioxide to zinc oxide. This will enable all the heavy powders to possess the same amount of slip although varying widely in covering power. The substitution in part or whole of titanium dioxide for zinc oxide in heavy powders yields the same degree of covering power as an increased percentage of zinc oxide content in heavy powders shows a tendency to cake.

The covering power of Titanium dioxide is about five times that of zinc oxide, which means achieving the desired effect with one fifth of the volume of zinc oxide used. However, because of the fact that it may not hold the colour as well as zinc oxide under the influence of the secretion of the face, it is advisable to mix zinc oxide.

Furthermore, as titanium dioxide does not dry as readily as zinc sulphide or zinc oxide, it is recommended that it be added to chalk in making the colour base with thorough milling.

Cold cream and mineral oil are added to face powder by some manufacturers, but it is difficult to say whether they serve any purpose, other than improving the adhesiveness of a poorly formulated face powder. In the case of cold cream, it is superfluous because the water phase evaporates leaving behind only the oil.

It is believed by some that magnesium carbonate adsorbes better then precipitated chalk and retains the perfume better. If both are used in the same formula, their ratio must be 2 : 1, and generally, the perfume oil is mixed with magnesium carbonate and the colour with chalk.

Increasing the stearate in the above indicated quantities leads to transparency of the powder after application.

RAW MATERIALS

It is of utmost importance to note that the quality of raw materials used for the manufacture of face powder should be of finest, whitest and highest quality.

The following grades of raw materials are recommended:

Zinc oxide	:	Finest, white grade.
Talc	:	Finest white, 99% through 200 mesh.
Zinc stearate	:	Whitest grade made from triple pressed stearic acid and odour free.

Precipitated chalk	:	Lightest, whitest grade.
Magnesium carbonate	:	Lightest, whitest grade.
Titanium dioxide	:	Finest white 200 mesh or better

MANUFACTURING METHODS

The perfume compound is mixed with a part of magnesium carbonate in an enameled white pail or a suitable vessel. The mixture is rubbed through a hand sieve with a stiff bristle brush once the oil is absorbed or it is run through a power brush sifter until the perfume is uniformly distributed. Then the colours are mixed in the same way with the rest of magnesium carbonate and chalk. The process is continued till no colour flakes show on a white paper when the mixture is rubbed out on it. A sample is kept aside for matching purpose.

The colour and perfume bases are then added to the rest of the raw materials into a mixer and mixed until they become uniform. It should be checked if it matches with the matching sample. Finally the powder is sifted at least to a 300 mesh product. Some manufacturers sift all the materials first before mixing in order to save cleaning of sieves.

Fineness of particle size has been described by terms like air floated, micronized and air spun, the process by which this condition is obtained. In air floating, the finished powder is passed through a mill equipped with a fan cyclone and a dust arrester to air separate the coarse particle to a predetermined height because they cannot be blown or floated, until they are adequately fine. A micronizer is a mill that grinds all the powder particles to a desired micron size (0.001 mm). The air spinning process employs a method wherein the powder is whirled around by a purified continuous air stream under great pressure. It is housed in a specially constructed cylindrical vessel. The powder particles knock against each other at an estimated speed of over thousand miles per hour. This collision at high speed reduces the particle size. At this point the smaller particles are sucked out through the use of a centrifugal force principle while the larger ones remain inside until they are divided further.

Processes like these render uniform distribution of perfume and colour and provide greater fluffiness. Some manufacturers in fact leave the finished powder in air tight bins for several weeks before filling it.

There is however, a second method of manufacture which employs two operations. First the preparation of a white powder base which is perfumed and stored in air tight tins to bloom. Secondly, sufficient quantities of colour bases are made at one time for several batches. This process speeds up the manufacture of face powder and also offers uniformity.

Formulas

White Base for Light Powders with Zinc Oxide

Zinc oxide	22.83
Talc	67.73
Zinc Stearate	6.25
Precipated chalk or Magnesium carbonate	2.25
Perfume	1.04
	100.10

White Base for Medium Powders with Zinc Oxide

Zinc oxide	26.04
Talc	65.46
Zinc stearate	5.21
Precipated chalk or Magnesium carbonate	2.25
Perfume	1.04
	100.00

With Titanium Dioxide

Titanium Oxide	3.00
Talc	88.50
Zinc stearate	5.21
Precipated chalk or Magnesium carbonate	2.25
Perfume	1.04
	100.00

White Base for Heavy Powders with Zinc Oxide and Titanium Dioxide

Titanium dioxide	3.00
Zinc oxide	21.25
Talc	68.29
Zinc stearate	4.17
Precipated chalk or Magnesium carbonate	2.25
Perfume	1.04
	100.00

Powder Bases

Peach (Also called nude, natural)

Precipitated chalk	59.0
Golden ochre	40.0
Brilliant pink lake	1.0
	100.00

Use 5 parts and 95 parts white powder.

Rachel

Precipitated chalk	75.0
Golden ochre	25.0
	100.00

Use 4 parts and 96 parts white powder

Flesh

Precipitated chalk	90.0
Golden ochre	5.0
Brilliant pink lake	5.0
	100.00

Use 4 parts and 96 parts white powder

Naturelle

Precipitated chalk	75.0
Golden ochre	24.0
Brilliant pink lake	1.0
	100.00

Use 4 parts and 96 parts white powder

Ochre

Precipitated chalk	54.0
Golden ochre	3.0
Brilliant pink lake	43.0
	100.00

Use 8 parts and 92 parts white powder

Suntan

Precipitated chalk	36.0
Golden ochre	58.0
Brilliant pink lake	6.0
	100.00

Use 9 parts and 91 parts white powder

ADDITIONAL FORMULAS

Talc	40.0	56.0
Colloidal clay	20.0	20.0
Zinc oxide	20.0	10.0
Zinc stearate	—	10.0
Precipitated chalk	15.0	—
Magnesium carbonate	5.0	4.0
	100.0	**100.0**
Talc	75.0	—
Zinc oxide	20.0	—
Zinc stearate	5.0	—
	100.0	
Talc	60.0	80.0
Zinc oxide	10.0	15.0
Zinc stearate	20.0	4.0
Magnesium carbonate	10.0	1.0
	100.0	**100.0**
Talc	69.0	82.79
Zinc oxide	18.0	—
Titanium dioxide	—	5.21
Zinc stearate	6.0	5.0
Precipitated chalk	7.0	7.0
	100.0	**100.0**

Talc	74.5	30.0
Zinc oxide	10.0	20.0
Titanium dioxide	10.5	5.0
Colloidal clay	—	40.0
Precipitated chalk	5.0	—
Magnesium stearate	—	5.0
	100.0	**100.0**
Zinc stearate	6.0	—
Talc	40.0	15.0
Colloidal clay	45.0	25.0
Titanium dioxide	4.0	5.0
Magnesium stearate	—	10.0
Magnesium carbonate	5.0	—
Precipitated chalk	—	45.0
	100.0	**100.0**
Talc	20.0	38.0
Colloidal clay	20.0	10.0
Zinc oxide	15.0	12.0
Magnesium stearate	5.0	13.0
Magnesium carbonate	10.0	2.0
Precipitated chalk	30.0	25.0
	100.0	**100.0**
Talc	15.0	67.0
Colloidal clay	35.0	20.0
Titanium dioxide	10.0	4.0
Magnesium stearate	5.0	3.0
Magnesium carbonate	5.0	6.0
Rice starch	30.0	—
	100.0	**100.0**
Talc	52.5	67.0
Colloidal clay	15.0	—
Titanium dioxide	—	3.0
Zinc oxide	12.0	20.0
Zinc stearate	7.0	4.0
Magnesium carbonate	2.5	—
Precipitated chalk	11.0	6.0
	100.0	**100.0**

Talc	51.5	51.5
Titanium dioxide	3.0	5.0
Zinc oxide	12.0	12.0
Colloidal clay	13.0	11.0
Magnesium carbonate	2.5	2.5
Precipitated chalk	11.0	11.0
Zinc stearate	7.0	7.0
	100.0	**100.0**

Talc	31.5	46.8
Colloidal clay	31.5	18.0
Zinc oxide	20.0	13.5
Titanium dioxide	—	4.5
Zinc stearate	8.0	8.1
Precipitated chalk	6.0	4.5
Magnesium carbonate	1.0	4.5
Mineral oil	2.0	0.1
	100.0	**100.0**

VARITIES IN FACE POWDERS

Compact powders, cake make-up powders, cream powders and liquid powders are closely related to face powders in composition and that is why we intend to discuss them at this point. Cosmetic stockings are similar to liquid powders and they are also being discussed here.

As we have already seen slip, adhesion, covering power, colour and odour are important properties of a good commercial face powder. This is equally true of compact powder which can be considered as a face powder moulded into a tablet. The processes involved in compact powders are very similar to those of tablet making in the pharmaceutical field. Methods of manufacture of cake rouge and compact powder are essentially the same, since rouge is merely a more highly coloured tablet of compact powder. The equipment used is also the same.

Wet compression method, dry compression method and wet moulding method, are the three common processes used for tablet face powder.

Wet Compression Method

The powder ingredients are mixed thoroughly just as in the manufacture of face powder. Fine powders cannot be readily compressed into a tablet. Further processing is necessary to form granules which is done by melting down with a liquid and binder and thorough mixing until a pasty mass is obtained. It is granulated by passing the damp mass through 1/ 8th inch mesh screen tacked over a frame through a specially built granulator. The wet granules are spread on trays and dried. Dried granules are then compressed by a tablet or compact press.

Dry Compression Method

In this method the powder base colour and perfume are milled by a pulverizer or ball mill. It is then moistened by a binding solution and mixed thoroughly until granular. These granules are then compressed and the finished cakes are dried in a drying cabinet at about 140 °F.

Wet Moulding Method

In this method all the ingredients are made into a wet, heavy paste. The paste is poured or pressed by rolling into lubricated nickel moulds and is allowed to dry. Dextrin or gum arabic adhesive is painted over the surface of the rouge cakes. Then glass, porcelain or metal plates are pressed down on the glued surface of the rouge cakes. When the cakes are dry they adhere to the plates.

A compact must be neither too hard nor too soft. The powder should come off easily on to the puff and the cake must not get hard and shiny. The basic ingredients used are the same as in face powder. The colours are mineral pigments, lakes and FD&C colours which are also similar to those used in face powder. The stearates make good binders and spreading agents. However, talc should be restricted to 50%, as more will cause cracking of cakes. The binding solution must be made in bulk quantities and the viscosity standardized. The binding solution is made from materials like gelatin, gum tragacanth, gum acacia, gum karaya, methyl cellulose, quince seed, rosin, Irish moss and occasionally lanolin dissolved in ether is added. The binding solution requires a preservative.

A typical binding solution formula is given below:

	Parts or %
Gum tragacanth mucilage (2%)	20.0
Quince seed mucilage (2%)	10.0
Gelatin mucilage (3%)	10.0
Rosin tincture	1.0
Water	58.8
Methyl *p*-hydroxy benzoate	0.2
	100.0

A typical base powder formula is

	Parts or %
Talc (300 mesh)	40.0
Zinc oxide	20.0
Zinc stearate	6.0
Rice starch	10.0
Magnesium carbonate	3.0
Colloidal clay	10.0
Colour pigment	10.5
Perfume	0.5
	100.0

To make a finished compact from these mix all the ingredients of the base formula together and run through a pulverizer, or a ball mill. Moisten the base powder with a sufficient quantity of the binding solution to make a suitable granulation for the type of process being followed, compress and dry the cakes.

Other formula for base powder follows:

	Parts or %
Talc California	62.0
China clay	15.5
Zinc oxide	7.5
Zinc stearate	7.5
Gum Arabic mucilage q.s. to granulate	
Perfume	
Lake colour	7.5
	100.0

Compact Powder Base No. 1

	% or Parts
Kaolin	40
Talc	40
Magnesium carbonate	10
Rice starch	10
Perfume	q.s
Anhydrous lanolin	q.s
Ether	q.s

Mix the powdered materials in a mixer together with the proper amount of lake colours. Granulate with the lanolin ether solution and binding solution.

Cake Make-up

In the manufacture and formulation of cake make-up, great care must be exercised to keep the formulas and processing standardized. The mixing operations, drying temperatures and fineness of powders must be kept uniform, else the result will be a substandard product. The use of cake make-up produces a flat, smooth, lasting finish to the skin, which is not achieved by any other cosmetic. It also conceals minor skin defects.

The composition of cake make-ups consists of ingredients used in face powder like talc, chalk, kaolin colloidal clay, titanium dioxide and zinc oxide, besides light or heavy mineral oils, vegetable oils, pigments, perfumes, water, humectants like glycerol and glycols, binding and emulsifying agents.

A well-formulated cake will come out easily with a moistened tissue or sponge as an emulsion and should cover the skin uniformly. Some products when still moist are blended into the skin with finger tips or cleansing tissues and others are left to dry on the skin. Whatever may be the technique of application, the film produced should not "draw" the skin by drying out quickly, remain on the skin throughout the day, repel moisture caused by perspiration, and be easily removed by washing with soap and water.

In order to obtain these properties the study of certain mineral earths will help us.

Titanium dioxide & zinc oxide: Impart covering and masking properties to the cake.

Kaolin and colloidal clay: Help as binders in compressing but excessive usage results in absorption of too much of water causing the film to pile up and become uneven.

Chalk: Regulates easy brushing off or blending with skin.

Talc: Is stable filler. But if used in excess shine will be imparted.

It is imperative for the chemist to combine these mixed earths on the other face powder ingredients judiciously to obtain the desired effect.

Coming to other ingredients, the pigments selected should not "bleed" during perspiration. Normally water insoluble lakes and mineral pigments are preferred. The oils used may be light or heavy mineral oils or vegetable oils and their function is to provide the desired oiliness. Care should be taken to prevent rancidity in vegetable oils by adding anti-oxidants. Further, the right choice of emulsifiers and right proportions must be used with mixed powders so as to prevent degreasing of skin.

The manufacturing process of cake make up differs when compared to compacts. First the powders are mixed. The water–oil emulsion and humectant are subsequently added and the mixture is passed through a roller mixer for greater homogeneity. The resultant paste is then granulated and pressed into cakes.

Liquid Cream Powders or "Night Whites"

Liquid and cream powders are often called "night whites" and are used for evening wear to counter the glare of electric lights. They are applied to face, neck and arms and serve to blend the colours of the skin exposed only in the evening dress due to their high opacity.

No. 1 Liquid Powder

Colloidal clay	18.0
Titanium dioxide	2.0
Glycerin	8.0
Water	71.5
Perfume	0.5
	100.0

No. 2 Liquid Powder

Colloidal clay	8.0
Precipitated chalk	5.0
Zinc oxide	10.0
Glycerin	5.0
Alcohol	5.0
Orange flower water	67.0
	100.0

No. 3 Liquid Powder

Talc	10.0
Colloidal clay	5.0
Titanium dioxide	5.0
Glycerin	10.0
Rose Water	64.5
Alcohol	5.5
Perfume	q.s
	100.0

No. 4 Liquid Powder

Colloidal zinc oxide	6.0
Precipitated chalk	8.0
Colloidal clay	3.0
Zinc stearate	2.0
Glycerin	3.0
Witch hazel	10.0
Orange flower water	68.0
	100.0

No. 5 Liquid Powder

Titanium dioxide	5.0
Precipitated chalk	8.0
Colloidal clay	5.0
Glycerin	3.0
Rose Water	79.0
	100.0

No. 6 Liquid Powder

Colloidal zinc oxide	8.0
Colloidal clay	5.0
Precipitated chalk, heavy	6.0
Glycerin	5.0
Witch hazel	10.0
Orange flower water	66.0
	100.0

No. 7 Liquid Powder

	% or Parts
Heavy mineral oil	55.0
In which is dissolved by agitation and heat	
Magnesium oleate	2.5
Titanium dioxide	25.0
Light chalk	3.75
Talc	6.25
Perfume	2.25
Red iron oxide	3.0
Light ochre	2.25
	100.0

Stir all other ingredients into the oil solution of magnesium oleate and allow to stand overnight before filling. In addition to the materials used here, barium sulphate, bismuth subnitrate, bismuth subcarbonate, lithopone and similar materials are sometimes employed. Some manufacturers add small quantities of starch or saponin in order to enhance adherence and stability. However, these additives must be carefully used so that the product is neither too sticky nor slow in drying out.

Manufacturing Methods

The powdered ingredients and colour, if any is used, are mixed together in a powder mixer. The liquid ingredients on the other hand are blended in a tank preferably fitted with an agitator. The powder is slowly introduced into the blended liquids with agitation. Once the powder is completely added, the mixture is stirred for half an hour. The finished product is filled with the stirrer in motion to insure uniform distribution of the powder in the liquid phase of the preparation, until the tank is emptied.

No. 8 Cream Powder

Vanishing cream	70.0
Talc	24.0
Titanium dioxide	5.5
Perfume	0.5
	100.0
Colour to suit	

Make the vanishing cream in the usual way. Mix the talc and titanium dioxide and colour and perfume. Add the cream and then run the entire mass over a roller mill.

No. 9 Cream Powder

Glyceryl monostearate	10.0
Glycerin	3.0
Heavy mineral oil	5.0
Spermaceti	5.0
Stearic acid	2.0
Caustic potash U.S.P	0.1
Water	48.4
Perfume	0.5
Titanium dioxide	6.0
Talc	20.0
	100.0

Dissolve the caustic potash in water then add all the rest of the ingredients with the exception of the perfume, but include the colour. Bring the mixture to a boil with constant stirring. Continue stirring until all the materials have melted and have become homogeneous. Shut off the heat and continue stirring until mass is cooled, then add the perfume. Run the mass through a roller of ointment mill if the colour does not come out uniformly.

No. 10 Cream Powder

Glycerin	42.0
Stearic acid	10.0
White face powder	42.0
Distilled water	4.0
Potassium hydroxide U.S.P	1.5
Perfume	0.5
	100.0

Colour to suit

Dissolve the potassium hydroxide in water, melt the stearic acid and add the caustic solution. Mix until saponification is complete. Heat the glycerin; add the face powder and colour and mix until all lumps have disappeared. Then incorporate the first mixture, add the perfume and mix again. Run the product through an ointment mill to ensure smoothness and uniformity of colour distribution. Lake colours are best for this type of product.

TOILET POWDERS

Toilet powders comprise of talcum powder, dusting powder or body powder or after bath powder, after shave powder and baby powder.

Talcum powder is the most important of these powders. As the name implies, it mainly consists of talc and other ingredients. They may not be used in the same large proportions as already discussed under face powders because users of talcum powder are looking for the characteristic slip of talc. Perfume and colour have large bearing on the sales of talcum powder, although white and natural shades sell the most. In some cases, camphor or boric acid are added; in such cases where medicinal claims are made by the manufacturers, such a powder falls under the category of drugs and not cosmetics.

Dusting powder, after shave powder, and baby powder are closely related to talcum powder. They have more or less the same composition as talcum powder. Dusting powder is sold with a puff to apply the same on person. After shave powder consists of talc with the addition of colour and other mineral ingredients so as to enable it to go on smoothly, cling to the face with less sheen and match the colour of skin. Baby powders are normally less heavily perfumed (for obvious reasons) and coloured. Boric acid because of its soothing and slightly antiseptic properties, is the favorite choice not only in baby powders but all toilet powders.

The manufacturing method for toilet powders is generally the same as face powders.

The following are formulas of talcum, body, and after-shave powders.

Talcum Powders

	No. 1
Talc	71.0
Precipitated chalk	20.0
Zinc stearate	3.0
Boric acid	5.0
Perfume	1.0
	100.0

	No. 2
Talc	54.0
Precipitated chalk	40.0
Zinc stearate	3.0
Boric acid	2.0
Perfume	1.0
	100.0

	No. 3
Talc	79.0
Magnesium carbonate	15.0
Boric acid	3.0
Magnesium stearate	2.0
Perfume	1.0
	100.0

	No. 4
Talc	19.0
Calcium carbonate	60.0
Rice starch	15.0
Boric acid	5.0
Perfume	1.0
	100.0

After Shave Powder

	No. 5
Talc	71.5
Titanium dioxide	3.0
Zinc stearate	4.0
Precipitated chalk	20.0
Golden ochre	0.5
Perfume	1.0
	100.0

Body Powder

	No. 6
Talc	50.0
Kaolin	30.0
Boric acid	2.0
Precipitated chalk	17.0
Perfume	1.0
	100.0

	No. 7
Talc	50.0
Precipitated chalk	25.0
Magnesium carbonate	20.0
Zinc stearate	2.0
Boric acid	2.0
Perfume	1.0
	100.0

	No. 8
Talc	68.0
Precipitated chalk	15.0
Colloidal clay	5.0
Boric acid	3.0
Magnesium stearate	3.0
Magnesium carbonate	5.0
Perfume	1.0
	100.0

	No. 9
Talc	70.0
Kaolin	13.0
Magnesium stearate	8.0
Precipitated chalk	5.0
Magnesium carbonate	3.0
Perfume	1.0
	100.0

	No. 10
Talc	70.0
Colloidal clay	11.0
Precipitated chalk	10.0
Zinc stearate	5.0
Boric acid	3.0
Perfume	1.0
	100.0

Baby Powders

	No. 11
Talc	63.63
Kaolin	20.0
Zinc stearate	5.0
Precipitated chalk	5.0
Boric acid	6.0
Oxyquinoline benzoate	0.12
Perfume	0.25
	100.0

	No. 12
Talc	66.75
Magnesium carbonate	5.0
Colloidal clay	10.0
Magnesium stearate	5.0
Boric acid	10.0
Titanium dioxide	3.0
Perfume	0.25
	100.0

	No. 13
Talc	54.0
Kaolin	20.0
Zinc stearate	5.0
Precipitated chalk	11.0
Boric acid	10.0
	100.0

There are other cosmetic powders such as deodorant powders, foot powders, hair powders, suntan preventive powders, and tooth powders, which can be covered under toilet powders. But as these products have special properties they are categorized and dealt with separately.

2

Hand Creams and Lotions

THE PRESENT DAY cosmetic market has many forms of hand products. Gone are the days of glycerine and rose water formulations. The most popular among the hand products are hand cream and hand lotion.

The popularity of these products relates to their genuine need—for either prevention of a rough, dry skin or treatment of such condition. Whatever the need may be, they are time-tested and have come to be a part of a cosmetic consumer's life.

The most common cause of roughened dry skin is generally accepted as long repeated contact with water alone or detergent solutions, particularly in cold water.

Through a series of fundamental experiments conducted by Blank it was established that water content of the stratum corneum is all-important in the maintenance of normal, soft, flexible skin.

Now, the treatment of dry skin involves the use of basic ingredients identified as emollients. It may be more or less defined as *an agent which, when applied to a dry or inflexible corneum, will effect a softening of that tissue by inducing rehydration.* This would immediately lead us to think of water as an ideal emollient. But the difficulty here is in its application. Only a thin film of water can be retained on the skin. As a result of which evaporation takes place before the emollient effect can be produced. The other possibility is to immerse the hands in water. However, this would tend to produce excess hydration leading to swelling of stratum corneum with the possibility of skin-cell damage.

The ideal emollient then must therefore be a substance that not only provides water to the stratum corneum but at the same time regulates its requirements.

Hence the formulation of cosmetic hand treatment products centre around an emollient along with other ingredients combined judiciously to evolve elegant hand creams and lotions.

Emollient cosmetic hand creams and lotions are somewhat unique in the sense that the solids used (which are ultimately deposited on the skin) have a higher melting point than the body temperature. This explains the relatively dry non-greasy feel imparted after application of these products. Moreover they create a "vanishing" effect when rubbed on the skin due to the dry film on the hand and the rapidity of water evaporation.

The conventional hand cream formula is a modified vanishing cream of the O/W type, the basic composition being a stearic acid soap as the emulsifier, an excess of stearic acid and a humectant such as glycerol and a high percentage of water. The hand lotion formula can be very similar; the difference being only the proportion of total solids.

Triethanolamine stearate is preferred over potassium or sodium stearate in the stearate type lotion, since it is a softer and more soluble soap than the other two. Though hand creams and lotions are formulated to achieve the same end-results, they differ in the usage of emollients while manufacturing.

Anionic emulsifiers, such as soaps have been used for long in hand creams and lotions. But the recent trend has been towards using nonionic and cationic emulsifiers to make non-soap cream and lotions, slightly acidic in nature, to match the acidity of normal skin which has an average pH of 5.5.

BASIC INGREDIENTS

Since the characteristics of a hand treatment formula are dependent upon the ingredients, it would be worth classifying and discussing these agents in some detail. The following ingredients are used in hand cream formulation:

1. Emollients 2. Barrier agents 3. Healing agents
4. Humectants 5. Emulsifiers 6. Preservatives
7. Perfume oil and 8. Colouring agents

Emollients

Agents like lanolin and its derivatives, sterols, phospholipids, hydrocarbons, fatty acids, esters and fatty alcohols are specifically used for their skin softening property.

Lanolin

Lanolin is widely used in hand treatment products. According to studies lanolin helps in maintaining the epidermis in a normal condition. Although disputed, lanolin according to these studies hardly penetrates the skin and is retained on the skin. However, its emollient properties are universally accepted.

Lanolin is a natural wax consisting of mainly esters formed by the union of higher alcohols and fatty acids. Its hydrophobic and adhesive character makes it an excellent occlusive agent and hence a good emollient.

The proportion of lanolin used does not generally exceed 5% because when used in higher concentrations it tends to impart "tackiness" to the end product.

Apart from being a good emollient it serves as a good emulsifier too and forms W/O type emulsions.

Lanolin Alcohols

Lanolin alcohols, derived from lanolin are commercially available and enjoy successful use in hand creams and lotions. These alcohols fall into three basic categories, sterols, triterpene alcohols and aliphatic alcohols.

The emollient effect produced by lanolin is generally attributed to the hydrophilic nature of sterols (sterol content of lanolin alcohols being 30%) perhaps more specifically to the cholesterol. The lipids present in the skin surface fat such as cholesterol are considered to be hydrophilic and not hydrophobic in contrast to general belief and upon prolonged contact allow soaking of water to a considerable degree. This rehydration of the corneum is due to the presence of cholesterol in the occlusive base.

Lanolin alcohols are available commercially in several forms.

1. **Solid waxy materials:** Yellow to amber colour.
2. **Liquid Solution:** Pale to golden yellow in colour. It is usually a solution of lanolin alcohols in mineral oil. Some outstanding examples of such a product are Amerchol L-101 and Nimlesterol. The first being a remarkable surface active agent apart from being an excellent emollient and the second is an excellent emulsifier and emollient.
3. **Unctuous Base:** The lanolin alcohols, being excellent emulsifiers themselves, have been combined with lanolin and hydrocarbons such as petrolatum, mineral oil and paraffin to produce ointment bases which when mixed with water yielded W/O emulsions spontaneously. The products are referred to as absorption bases and are commercially available. Some of these bases are modified with the addition of emulsifiers such as glycol esters, ethylene glycol esters, and sorbitol esters.

Modified Lanolin

The following lanolin modifications which are advantageous in the product development should be considered as far as the hand treatment products are concerned.

Liquid Lanolins

Liquid lanolin is obtained by fractionating lanolin. This liquid form of lanolin is an effective emollient and enjoys the following advantages over lanolin in hand cosmetics

1. Appreciably less drag and stickiness.
2. Considerably greater solubility in hydrocarbons, even at low temperatures.
3. Greater concentration feasible in hand creams and lotions.
4. Greater ease of handling.

Esterified Lanolin Alcohols

These lanolin esters have the same physical appearance as natural lanolin but are appreciably soluble in mineral oil unlike lanolin. When used in hand creams and lotions they deposit on the skin films that are hydrophobic, waxy, protective and emollient in character. However, one important point concerning these esterified lanolins is that they do not possess the emulsifying property that is so characteristic of natural lanolin.

Acetylated Lanolin Alcohol

This is an emollient which has been developed as a liquid fraction of acetylated lanolin alcohols. When used in hand creams and lotions it is deposited on the skin as an emollient that was found to be exceptionally nontacky and extremely hydrophobic.

Polyoxyalkylene Lanolins

The idea behind the creation of these lanolins was to incorporate natural lanolin, an excellent emollient by itself although hydrophobic in nature, with hydrophilic properties. They were produced by reacting ethylene oxide and lanolin and are semi-solid in nature, somewhat less viscous than natural lanolin, and more soluble in water (than lanolin). This solubility varies directly with the ethylene oxide chain length.

Polyoxyalkylene lanolins exhibit the following advantages in hand cream and lotions formulations.

1. They deposit on the skin an occlusive film that is non insulating.

2. They are emollient without excessive stickiness.
3. They can be used as primary emulsifiers, yielding oil-in-water emulsions.
4. They impart greater surface activity.
5. They are excellent hydrophilic plasticizers in creams with high solid content.

Alcohol Lanolin Ester

Another lanolin derivative is the product obtained from partial transesterification of lanolin with isopropanol. It usually contains 30%–50% residual unreacted lanolin.

It lends to hand creams and lotions the solubilizing and plasticizing features of isopropyl alcohol apart from being an emollient.

Sterols

Cholesterol is the only sterol, which appears to have a specific application as an emollient to hand creams and lotions and the reason why it is so may be linked to the facts listed hereunder.

1. The surface of the skin is covered by a greasy layer consisting mainly of waxes, free and esterified cholesterol.
2. An analysis of the skin surface fat shows 2.5% free cholesterol and 2.5% esters of cholesterol.
3. The lipids present in the skin-surface fat, such as cholesterol, hydrophilic upon prolonged contact, enable the layers of the skin to take up water to a considerable degree.
4. Dermatologists have recommended the use of cholesterol to lessen the irritating and defatting action of soap.
5. It is maintained that the sterols present even in an emulsified form for instance in soap lather, easily penetrate into the epidermis and impart suppleness to skin.

Phospholipids

Phospholipids are complex fat-soluble substances that contain in their molecule a nitrogenous base (such as choline or ethanolamine) in addition to fatty acids and glycerol.

Lecithin, a phospholipid, has found a place in the hand cream and lotion formulations as an emollient. Its concentration does not generally exceed 5%; infact, it usually hovers around the 1–2% mark. Lecithin is not only an emollient but is an excellent emulsifier as well as a surfactant.

Hydrocarbons

The hydrocarbons like petrolatum, mineral oil, paraffin wax and ozokerite have been used in emollients in hand creams and lotions only to a certain effect. This is because they impart an uncomfortable feeling of warmth to the skin and also leave a sticky or waxy effect.

Fatty Acids

Fatty acids have earned the distinction of being one of the essential ingredients in the formulation of hand creams and lotions. Of these stearic acid is the only one of choice.

Stearic acid figures in most hand creams and lotions. In some as part of the emulsifier (such as potassium stearate or triethanolamine stearate) but in the majority of cases it is also present as the free acid with proportions varying from 1–20% depending on the consistency of the finished product.

Stearic Acid

The percentage of stearic acid used in hand creams being appreciable, the consistency of the end product depends greatly on the type used.

The important fundamental difference between one type of stearic acid and the other is in the method used in its manufacture. Here are some of the various types of methods employed in the manufacture of stearic acid.

Stearic Acid obtained by Pressing

The fatty acid mixture obtained from splitting of tallow is separated into a liquid and solid portion by a series of pressings. The solid phase is made up of saturated fatty acids with higher molecular weight like stearic and palmitic. Commercial "triple pressed" stearic acid generally consists of a combination of 55% palmitic and 45% stearic acids. The liquid phase consists of oleic, linolic, and myristic acids. The three grades of stearic acid: triple pressed, double pressed and single pressed represent the pressing during separation.

In hand cosmetics the triple pressed grade is generally favoured, sometimes double is employed where pearly sheen and softer consistency are desired. The single pressed is hardly used because of its high percentage of unsaturated fatty acids which have a tendency to cause rancidity.

Stearic Acid obtained by Solvent Crystallization

Stearic acid obtained by this process involves fractional crystallization of solid fatty acids from solvent solution and the successive crystallization determines the degree of purity of the final product. A commercial grade similar to triple pressed stearic acid can be obtained by this method.

Stearic Acid obtained by Hydrogenation

Nearly pure fatty acids can be obtained by a combination of hydrogenation and fractional distillation. And commercial stearic acids with 97% stearic acid content are now available.

Stearic Acid obtained by Fractional Distillation

The fatty acids obtained after fat splitting may be separated into compounds with different chain lengths by fractional distillation. However, oleic and stearic having the same chain length are distilled at the same temperature and to obtain pure stearic acid first, a solvent crystallization or a pan press method is necessary to get rid of the oleic acid before fractional distillation.

With newer methods available for making commercial fatty acids it was possible to obtain stearic acids of different stearic content. Further, it was found in hand cream formulations that by substituting stearic acid with a different stearic content for the commercial triple pressed stearic acid, the consistency of the cream varied. For instance, reports of softening effects were recorded when high purity stearic acid was used in place of the conventional triple pressed stearic acid in the presence of nonionic emulsifiers.

Fatty Acids Esters

Fatty acids esters of low molecule weight have been popularly used in hand creams and lotions. Butyl stearate, isopropyl stearate, isopropyl palmitate, and isopropyl myristate are some products belonging to this group. Their oily character and low viscosity has made them desirable agents in hand product formulations. They leave a thin oily film on the skin which in nontacky or greasy. The nature of the films being hydrophobic and continous makes them good emollients. They are generally used at 2–10% concentrations in hand products.

A word of caution here regarding stearate type hand products: the isopropyl esters are to be used judiciously because they have a tendency to increase the gelation.

The polyol ester like glyceryl monostearate and propylene glycol monostearate, ethylene monostearate and polyethylene glycol monostearate, are emollients that have served in the development of modern hand creams and lotions. Except polyethylene glycol monostearate all the other agents produce a waxy, occlusive, water-insoluble films. Polyethylene glycol esters are hydrophilic; hence their films are not completely occlusive; however, they are still satisfactory emollients. In addition to being emollients, they are useful emulsifiers and it is for this very purpose they are used in hand product formulations.

Concentrations of glyceryl monostearate, propylene glycol monostearate, and ethylene glycol monostearate have a direct bearing on the viscosity of the emulsion. They are generally employed in concentration of 0.5–5% and 1–10% in hand creams and lotions respectively. The nature of polyethylene glycol esters is directly linked with ethylene oxide chain lengths. The longer the chain the more solid the product and also the more hydrophilic it becomes.

The higher alcohol esters of fatty acids, such as cetyl palmitate (or Spermaceti Wax), are quite waxy in nature and are considered excellent emollients. The occlusive waxy nature of these agents limits them to acceptable concentrations of about 0.5–2% and upto 5% in hand lotions and creams respectively.

Fatty Alcohols

The fatty alcohols have been widely used in hand creams and lotions. Cetyl and stearyl alcohols being the more favoured ones. Lauryl and myristyl have also been in use in hand product formulations but as a rule in combination with cetyl alcohol, stearyl alcohol or both.

As emollients cetyl and stearyl alcohols have been found to be very effective. Being hydrophobic they produce occlusive films that help in inducing hydration of skin. Further, they have sufficiently high melting points so as to deposit non-greasy films on the skin. A combination of cetyl and stearyl at concentrations as low as 0.2% of each are known to impart to the hands a smooth velvety feel.

To have an acceptable end product the concentration of cetyl and stearyl alcohol have to be very cautiously worked out particularly in soap systems, because the rate of occurrence of gelation is directly related to their concentration.

BARRIER AGENTS

There has been a growing interest among the hand-product customers ranging from a housewife to the garage mechanic who constantly risk irritation in their daily routines, in a group of emollients that in addition to promoting rehydration of the stratum corneum aid in protecting the skin surface. These ingredients are called "barrier agents."

Protective hand creams and lotions have recently increased in number and variety. There is a standing demand for such products, which provide protection against domestic and industrial materials that are likely to cause skin irritation.

A well-formulated barrier cream needs to fulfill the following basic requirements.

1. Good consistency and ease of application.
2. Good and reasonably persistent adherence to the skin.
3. Ability to form a coherent, impervious, flexible, and noncracking film.
4. Freedom from any tendency to irritate skin.
5. Ease of removal when desired.
6. Esthetic acceptability, from the point of comfort on application, relative imperceptivity, etc.

It would be ideal if all the above features could be incorporated into a single hand product. However, it would seem rather unlikely to expect a single product to assume complete protection since irritants vary considerably in physical and chemical properties.

Protective hand creams at present fall into two categories.

1. Water repellent.
2. Oil repellent.

The water repellent creams act as barriers to water and water-soluble irritants. The oil-repellent protective products act as barriers against oil and oil-soluble irritants.

There exist some products with a duel function. But they cannot be considered universally protective. Nevertheless they represent a step in that direction.

The advent of silicones offered a wide range of possibilities and combinations with some other better known barrier agents in formulating excellent protective hand creams and lotions.

Here is a list of cosmetic materials showing outstanding skin protective or barriers properties.

> Petrolatum, paraffin wax, ozokerite wax, vegetable waxes, beeswax, casein, methlycellulose, sodium carboxymethylcellulose, alginic acid salts and derivatives, zein, tragacanth, pectin, quince seed gel, bentonite or veegum, zinc stearate, sodium silicate, talc, stearic acid, titanium dioxide, and silicones.

HEALING AGENTS

The need to include healing agents in hand creams and lotions has been in the reckoning for a long time. The reason for this being the severe chapping of the hand leading to craking of the epidermis which could be very painful. Actually, it may be considered as a wound. Furthermore, the constant use of hands in the daily routine could leave them bruised and scratched. Some hand products contain agents that act as a skin healer and their function is to stimulate the growth of healthy tissue. Two basic agents have been identified as skin healers in urea and allantoin. These agents infact are chemically related, allantoin being a uric acid derivative.

Here are 5 basic attributes of allantoin.

1. Allantoin effectively "digests" tissue to produce a "natural" cleaning up of necrotic material.
2. It is a remarkable cell proliferant and quickly stimulates the development of healthy granulation tissue, thus reducing healing time.
3. These actions are accomplished without pain; in fact, pain frequently tends to be reduced, if not entirely relieved, when allantoin is applied.
4. It may be employed in dilute solutions and, therefore, does not dry out or cake, but remains in intimate contact with affected tissues.
5. It may be applied in solution, emulsion, or ointment form either alone or in combination with other therapeutic agents. Therefore, the addition of 0.01 to 0.1% to various cosmetic preparations, such as skin creams, tonics, lotions, soaps and shaving preparations, would enhance their healing properties.

Urea has been used in hand products. It is found to be beneficial when applied to infectious lesions. It is easily available and is non-toxic.

However, from a formulation point of view it should be noted that urea when used in its normal concentrations of 3–5% presented the problem of discoloration after a six-month ageing period.

HUMECTANTS

Humectants are agents that control the moisture exchange between the product and air, both in the container and on the skin. They have perhaps found more extensive use in hand creams and lotions than any other product. Even the earliest hand treatment product contained a mixture of 50% humectant and 50% water. There are many agents that possess the properties of a humectant; but only three have been widely used in hand creams and lotions, glycerol, propylene glycol and sorbitol. They are all organic compounds and are similar in that they are all polyhydric alcohols. However, they are different as far as molecular weight, viscosity and volatility are concerned.

Table below compares their physical properties quantitatively.

		Mol. Wt.	*Viscosity*	*Volatility*
1.	Propylene glycol	Lowest	Lowest	Highest
2.	Glycerol	In between	In between	In between
3.	Sorbitol	Highest	Highest	Non-volatile

Their behavioral patterns when used in hand creams and lotions too were dissimilar.

The effects of these agents on the rate of loss of water (in weight) in hand creams of O/W type were

1. In stearic acid soap-type hand creams, sorbitol in concentrations of 2 to 20% and at relative humidities of 30, 50, and 70%, retarded moisture loss more effectively than propylene glycol and glycerol.
2. In nonionic-type hand creams, the difference between the three humectants was less pronounced. They were all about equally effective in inhibiting moisture loss, although at 30% relative humidity and at 2, 5, 10, and 20% concentrations, propylene glycol was more effective than either sorbitol or glycerol.

During World War II when glycerin was scarce owing to its use for the military, other humectants were selected and substituted.

However, it was found that the choice of humectants could not be changed at will.

Here is a comparative study of effects of polyols on emulsions.

1. The consistency of an oil-in-water hand cream was related to the polyol used in the following manner: (a) glycerol produced creams having the hardest consistency; (b) sorbitol produced creams of medium hardness; and (c) propylene glycol produced the softest creams.
2. The consistency of an oil-in-water, soap-type hand lotion was related to the polyol used in the following manner; (a) glycerol produced lotion with the best flow characteristics; and (b) propylene glycol and sorbitol showed a tendency toward gelation.

It was found that the reason for this dissimilarity in their behavior could be attributed to the solubility of stearic acid in the various polyols rather than to their hygroscopic nature.

Humectants apart from being used for their hygroscopicity and emulsion consistency are excellent plasticizers. Infact the concentration of a polyol in a system is determined by the amount of solids that need to be plasticized. A cream or lotion properly plasticized prevents "rolling" or in other words will apply smoothly and uniformly.

The ability of humectant to release water gradually is best seen when the cream is supplied to the hands. The controlled loss of water from the emulsion permits a smooth inversion and prevents breaking of the emulsion which otherwise may result in an unacceptable "watery" feel.

Glycerol, propylene glycol, and sorbitol have another added advantage in that they are generally innocuous and have no side effects.

Other agents that qualify as humectants in cosmetics include polyethylene glycerols, mannitol, polyethylene sorbitols, and polyethylene glycols.

The need for humectants is almost indispensable in hand creams and lotions. However, its choice will depend on a number of factors some of which have been discussed.

EMULSIFIERS

The increasing knowledge about emulsifiers, which have made possible a successful union of water and oil, has been greatly influencial in enhancing the cosmetic elegance of hand products.

The emulsifiers used in hand creams and lotions are categorized into three types

1. Anionic,
2. Cationic, and
3. Nonionic

In this section we shall concern ourselves to know about outstanding emulsifiers used in hand creams and lotions and a brief description of their behavior in these products.

Anionics

This group of emulsifiers is widely used in the formulation of hand creams and lotions and account for about 75% of the hand products available in the market. Some examples of anionic emulsifiers are shown in the table below.

Anionic Emulsifiers

Type	Examples
Fatty acid soaps	Potassium stearate
	Sodium stearate
	Ammonium stearate
	Triethanolmine stearate
Polyol fatty acid monoesters containing fatty acid soaps	Glyceryl monostearate containing either Potassium or Sodium soap
Sulphuric ester (Sodium salts)	Sodium lauryl suphfate
	Sodium cetyl sulphate
Polyol fatty acid monosters containing sulphuric esters	Glyceryl monostearate containing Sodium lauryl sulphate.

The fatty acid soaps and sodium salts of sulfuric esters are strongly hydrophilic and will tend to produce O/W emulsions.

The polyol fatty acid monoester soaps are only slightly hydrophobic and will tend to produce dual emulsions. The most commonly used fatty acid soap in hand lotion is triethanolamine stearate in concentrations ranging from 0.5–3%. Fatty acid soaps are known to produce stable emulsions but on standing they tend to thicken and finally gel. This is a common feature seen in commercial hand lotions. Fatty alcohols and polyol fatty esters have even a greater tendency than fatty acid soaps and therefore caution is necessary when combining the two in formulation. Small quantities of sulphuric esters such as sodium lauryl sulphate prevents gel formation. Also an increase in the quantity of mineral oil (10–20%) and of polyols has a retarding effect on this gel formation.

In hand creams formulations that use only one emulsifier in sodium stearate, the cream produced is very hard initially. However, on standing it becomes softer and eventually "soapy". The reason for this being the insoluble nature of sodium stearate used in small proportions or in conjunction with more soluble stearates. Ammonium stearate and amine salts of fatty acids generally produce white creams, which tend to discolour towards yellow upon ageing. This is "Catalyzed" by the presence of trace metals, especially iron.

This tendency to discolour is also partially attributed to emulsification temperatures. Hence it is suggested that creams containing ammonium stearate or other amine salts, as emulsifiers should be produced at the lowest possible temperature.

Cationics

This group of emulsifiers has not been so widely used in hand creams and lotions. However, lately, these agents have found use in the production of presumably unique hand cosmetics. The use of cationics in hand creams can be summarized as

1. Being substantive to protein at an acid pH.
2. Instrumental in producing emulsions with an acid pH.
3. Being germicidal when not inactivated by anionic or other incompatible materials.

The cationics generally used in hand creams and lotions are listed below.

1. N (stearoyl colamnio formylmethyl) pyridinium chloride.
2. N-soya-methyl morpholinium ethosulphate.
3. Alkyl dimethyl benzyl ammonium chloride.
4. (Diisobutyl phenoxy ethoxy) ethyl dimethyl benzyl ammonium chloride.
5. Cetyl pyridinium chloride.

Emulsions produced by cationic emulsifiers are quite stable and show least tendency to gel. On the contrary they tend to thin out with age. A small amount of polyol fatty acid ester such as glyceryl monostearate helps to prevent this problem.

Cationic systems show novel possibilities in hand creams and lotions.

Nonionics

One basic difference between nonionic emulsifiers and the other two cationic and anionic is that the former shows no tendency to ionize. This property makes them compatible with other nonionics as well as electrolytes even in high concentration.

The reasons for using nonionics as emulsifiers in hand creams and lotions are

1. In hand creams, nonionic emulsifiers do not tend to produce a surface crust.
2. Oil-in-water hand creams made with nonionic emulsifiers show the least amount of shrinkage due to water evaporation.
3. Hand lotions and creams containing nonionic emulsifiers are extremely resistant to freezing.
4. Germicidal agents of the cationic type can be incorporated into hand creams and lotions without fear of incompatibility.
5. Acid, neutral, or alkaline hand creams and lotions can be readily formulated with nonionics.

Some of the nonionic emulsifiers used in hand creams and lotions are

Type	Examples
Polyoxyethylene fatty alcohol ethers	polyoxyethylene lauryl alcohol.
Polyoxyethylene fatty acid esters	polyoxyethylene stearate.
Polyoxyethylene sorbitan fatty acid esters	polyoxyethylene sorbitan monostearate.
Sorbitan fatty acid esters	sorbitan monostearate.
Polyoxyethylene glycol fatty acid esters	polyoxyethylene glycol mono and distearate
Polyol fatty acid esters	Glyceryl monostearate propylene glycol monostearate.

The use of nonionic emulsifiers in hand creams and lotions involves two fundamental facts.

1. A strongly hydrophobic ester must be used to emulsify the stearic acid or their oil or wax material. Outstanding examples of such agents are polyoxyethylene sorbitan fatty acid esters, such as sorbitan monostearate or the polyoxyethylene glycol fatty acid ester, such as polyoxyethylene glycol 1000 monostearate.

2. A slightly hydrophobic ester or lipophilic ester must be used in the same system in order to insure consistency of the end product. Examples of these nonionic can be found among the sorbitan fatty acid esters, the polyol fatty acid esters, and even the low molecular weight polyoxyethylene glycol fatty acid esters.

The proportions of each of these two types of emulsifiers varies considerably from about 1–10% of each in hand creams and 0.5–3% of each type in hand lotions. Although it must be noted in creams the concentrations of the two emulsifiers fall between 1–5%.

Apart from the emulsifiers discussed, there is another class of emulsifiers that has been used in hand products. They are represented by natural gums like (tragacanth and the alginates), the cellulose gums (e.g. methylcellulose and sodium carboxymethyl cellulose), the clays (e.g. bentonite and veegum), and the synthetic polymers (e.g. polyvinyl alcohol) which produced hydrophobic protective colloids. Although these agents have emulsifying properties, they are not primarily emulsifiers but fill in the role of emulsion and suspension stabilizers. They also serve as aqueous phase thickeners. When calculated on a dry basis, they were usually used in proportions under 1% in hand creams and lotions. These colloids were prepared as aqueous solutions or dispersions just before production since sufficient time is needed for their hydration.

PRESERVATIVES

Hand creams and lotions contain water and other ingredients, which are susceptible to attack by microorganisms. Hence the absolute necessity for a good preservative. The following are the properties for a good preservative.

1. It must be effective against all types of microorganisms causing decomposition.
2. It must be soluble internally or externally.
3. It must not be toxic internally or externally.
4. It must be compatible; must not alter the character of the preparation as far as objectionable odour, colour, taste, and other properties are concerned; and must be practically netural so that it will not alter the pH of the preparation.
5. Its cost should not increase the price of the preparation to any marked extent.
6. Its inhibiting effect must be lasting; therefore it may not be possible to depend on volatile substances, the effects of which disappear after evaporation.

Over the years many preservatives have been tried but have been discarded for one reason or the other. For example, benzoic acid, sodium benzoate, and sodium propionate have been successful in acid medium only; therefore they are of no use in creams and lotions that are slightly alkine in nature. Salicylic acid was considered, but discarded because of its skin irritation potential. However, the most ideal preservatives were found to be the esters of hydroxybenzoic acid. They were reported to have been 2–3 times more effective than benzoic acid in preventing bacterial growth.

The methyl, propyl, and butyl esters of hydroxybenzoic acid are generally used in hand creams and lotions preferably in combination with an other.

The most commonly used esters in hand creams and lotions are the methyl, propyl and butyl. A concentration of 0.1–0.2% methyl *p*-hydroxybenzoate is sufficient in hand lotions. If cholesterol or lecithins are found in appreciable quantities in the oil phase an additional 0.05% of propyl *p*-hydroxybenzoate can be added to this phase.

In hand creams where the oil content is generally high—0.05 to 0.25% of the methyl ester in the water phase and 0.05% of the propyl ester or butyl ester in the oil phase—are suggested.

Method of Incorporation of Preservative

The methyl ester is added in the water phase heated to 60°C with constant stirring during heating.

The propyl ester or butyl ester is added to the oil phase with constant stirring during heating.

These agents have shown most satisfactory results as preservatives in hand creams and lotions and their activity may be summarized as follows.

1. Antimicrobial studies show that the methyl, ethyl, propyl, and butyl esters are effective in low concentrations against fungi and grampositive bacteria, but less effective in gram-negative bacteria.
2. The esters are more fungistatic than fungicidal.
3. Their effect is additive, suggesting the use of combinations to achieve higher concentration in water.
4. They are substantially equally effective against microorganism in acid or netural solutions in the pH range of 4 to 8.

Perfumes

The choice of a perfume for hand creams and lotions is purely based on aesthetic value. The odour of perfume is one of the important factors in the overall acceptance of a hand product but one must not forget the importance of the compatibility of perfume oil with the emulsion besides ensuring that it is not a skin-irritant or sensitizer.

It must be noted that many essential oils, synthetic aromatic chemicals, and other perfume materials possess surfactant properties and can easily interfere with the performance of the basic emulsifiers. It was found that terpineol, hydroxycintronella, geraniol, eugenol, methyl eugenol, and phenyl acetaldehyde could severely effect the stability of emulsions with cationic or anionic emulsifiers. The effect of the perfume varies depending on the concentration and the emulsion system employed.

Study of the effects of perfume materials on the stability of cosmetic emulsions, specifically of certain perfume agents on several types of hand lotion system yielded the following results.

1. **Synthetic aromatics**: Terpineol extra, phenylethyl alcohol, geraniol pure, hydroxycitronella, and amylcinnamic aldehyde.
2. **Essential oils:** *Rose de mai* absolute, geranium bourbon, and lavender (50% ester content).
3. **Compound perfume oils:** A "medium" bouquet a lilac type, a light floral type, and two of a rose character.

The four hand lotion systems evaluated by us gave the following results:

1. **Triethanolamine stearate emulsion:** Of all the ingredients tested, terpineol was the only one to cause separation of this type of emulsion.
2. **Amino glycol emulsion:** A rose perfume composition in a 1% concentration caused this emulsion to separate.
3. **Potassium stearate–quince seed mucilage emulsion:** Geranium caused separation in all concentrations from 0.25

to 1 per cent. Terpineol extra caused separation in the 0.5% and 1% concentrations. A light floral bouquet caused separation only in the 1% concentration.

4. **Sorbitan monostearate–polyoxyethylene sorbitan monolaurate emulsion:** None of the materials tested affected the stability of this emulsion.

In stearate type creams, perfume ingredients like indole, methyl anthranilate, eugenol, ethyl vanillin, etc. to name a few, must be avoided because the pH is usually alkaline and discolouration occurs.

The difficulties encountered with perfume oils in hand creams and lotions cannot be seen until the product is stored for some time. Hence it is imperative to select the perfume wisely and conduct rapid ageing shelf-tests.

Another problem that manifests almost instantly is associated with the allergic hypersensitivity for which the standard patch test should be employed using the product on a large section of people.

COLOURING AGENTS

Colour is a psychological aspect of a product; people tend to accept or reject the same product when tinted with colour or left uncoloured. Hand creams and lotions are usually left uncoloured although a recent survey indicates that hand products in shades of pink and blue are more acceptable than those that are white and uncoloured.

The selection of colour by the formulator demands two aspects: one the psychological safety, and the other its compatibility with the product. In USA the psychological safety is simplified by restriction on use of coal tar colours only.

The more realistic problem is one concerning the compatibility of colour with the product. It is related to factors such as pH, solubility, and stability in the presence of light, metallic ions and oxidising and reducing agents.

Most hand creams and lotions being O/W emulsions, water-soluble colours should be used to tint the water, which is the external phase. In case of W/O emulsions, oil soluble colours are in use. Sometimes it may be advantageous to tint both phases of an emulsion in which case both water and oil soluble colours are employed in the same product.

The pH level of most hand creams and lotions lies between 5 and 8. Some of the later cationic systems are strongly acid, with pH 2. Some colours are stable in acid media but are very unstable in alkaline media. D&C Red No.20 is one such example. It is precipitated out of solution in alkaline media. The reverse is also possible. D&C Red No.23 is stable in alkaline conditions but tends to precipitate and also discolour towards yellow shade.

Some cations with high molecular weight react with anionic colours resulting in insoluble complexes, which precipitate in the presence of cations.

Hand lotions that are packed in clear glass bottles tend to fade on exposure to sunlight. Hence fastness to light is an important factor to be considered in colouring agents.

Some perfume oils contain reducing agents that cause the fading of hand product. This is observed with a number of red and blue certified colours. In the presence of reducing agent, and more so in alkaline media they will revert to their respective leuco form.

Some certified colours with good over all stability in hand creams and lotions are

RED (PINK): FD&C Red No. 1, FD&C Red No. 2, D&C Red No. 5, D&C Red No. 19, D&C Red No. 33.

BLUE: FD&C Blue No. 1, FD&C Blue No. 4, Ext D&C Blue No. 1, Ext D&C Blue No. 2, Ext D&C Blue No. 4.

Formulation and Manufacture

The formulation of hand creams and lotions depends on the purpose it is meant for. For example the formulator could expect to exhibit the following features.

1. Should soften the hands.
2. Should apply easily and quickly.
3. Should not leave a tacky film.
4. Should not interface with normal hand perspiration.

5. Should be antiseptic.
6. Should have a pleasant smell.
7. Should have a stable and appealing colour.

Keeping the above objectives in mind, he may also consider the points hereunder.

1. One or more emollients can be used for softening of the skin.
2. A vanishing cream base will accomplish the second objective. The use of alcohol might assist in imparting the apparent "vanishing" quality to a hand lotion formula.
3. The careful selection and proper combination of the waxes, oils and humectants will control the amount of tackiness developed by creams and lotions.
4. The judicious choice of the solid ingredients will aid in the prevention of excessive occlusion the fourth objective.
5. The choice of the antiseptic might well determine whether the emulsifier should be anionic, cationic or nonionic emulsifier.
6. The selection of pleasant perfume oil that is compatible with the emulsion and does not overpower the fragrance worn by the user will solve the sixth objective.
7. Choosing a stable colour will involve the consideration of such factors as pH, presence of reducing compounds, effects of light and type of emulsion.

By examining and discussing some typical formulas hereunder, the importance of practical consideration associated with both formulation and manufacture can be exposed to the reader.

The rate, time and type of stirring also play an important role in the stability and consistency of the cream. Hand cream systems, in general, are thixotropic. Therefore, excessive and vigorous shearing action reduces the viscosity beyond acceptable levels. Often the effect of prolonged and vigorous agitation means an irreversibly low viscosity.

Prolonged stirring results due to inefficient cooling. In summer when the cooling water is warm the cooling cycle automatically increases. This means it will take a longer time to arrive at finished filling temperature. Hence, to ensure smooth production, the cooling cycle must be established and maintained with the help of a refrigeration unit to keep the water at a constant temperature at all times. On occasion this type of stirring presents a different problem. For example, if the stirrer does not sweep the walls of the container clean of the solid deposits, a layer is formed which acts as an insulating medium thereby preventing efficient cooling. This cold solid mass also produces a grainy effect in the finished product when it is finally distributed throughout the cream. A sidewall scraper will effectively correct both these situations.

Formulas for Hand Creams

Part A

Cetyl alcohol	2%	—	10%	—	—	—	—
Glyceryl monostearate pure	—	—	—	—	—	10%	—
Isopropyl myristate	—	—	—	—	—	—	3%
Isopropyl palmitate	—	—	—	1%	3%	—	—
Lanolin	1	1%	—	—	—	2	—
Mineral oil	2	—	—	—	—	—	—
Polyethylene glycol 1000 monostearate	—	—	—	—	5	—	—
Polyethylene sorbitan monostearate	—	—	—	1.5	—	—	—
Propyl paraben	—	0.05	—	—	—	—	—
Sodium cetyl sulphate	—	—	2	—	—	—	—
Sorbitan monostearate	—	—	—	2	—	—	—
Stearic acid	13	16	8	15	20	—	17
Stearyl alcohol	—	—	3	—	—	—	—
Stearyl colamino formyl methyl pyridinium chloride	—	—	—	—	—	1.5	—

Part B

Glycerol	12	—	8	—	—	15	10
Lauryl colamino formyl methyl pyridinium chloride	—	—	—	—	—	—	5
Methyl paraben	0.15	0.15	0.1	0.1	0.15	0.1	0.1
Polyethylene glycol 300 monostearate	—	—	—	—	—	—	—
Propylene glycol	—	10	—	—	—	—	—
Potassium hydroxide	1	0.6	—	—	—	—	—
Quince seed mulcage (2% solution)	—	25	—	—	—	—	—
Sodium lauryl sulphate	—	—	1	—	—	—	—
Sorbitol	—	—	—	3.5	3	—	—
Triethanolamine	—	0.3	—	—	—	—	—
Water	68.85	46.9	67.9	76.9	68.85	71.4	64.9

Formulas for Hand Lotions

Part A

Cetyl alcohol	0.5%	0.5%	—	—	—	—	1.5%
Glyceryl monostearate	—	—	1%	4%	—	1%	—
Isopropyl palmitate	—	—	4	—	—	3	—
Lanolin	—	—	—	—	1%	1	—
Lanolin absorption base	—	—	—	1	—	—	—
Mineral oil	—	—	—	—	—	—	3
Polyethylene glycol 400 distearate	—	—	2	—	—	—	—
Propylene glycol monostearate	—	—	—	—	4	—	—
Nimlesterol or Amerchol L-101	—	—	—	7	—	—
Stearic acid	3	5	1.5	—	—	2

Part B

Glycerol	2	2	10	3	—	5	7
Methyl paraben	0.1	0.1	0.1	0.1	0.1	0.1	0.1
N (lauroyl colamino formyl methyl pyridinium chloride)	—	—	—	—	—	—	1.5
Propylene glycol	—	—	—	—	3	—	—
Sodium alginate	—	0.3	—	—	—	—	—
Sodium cetyl sulphate	—	—	5	—	—	—	—
Sodium lauryl sulphate	—	—	—	1	—	—	—
Stearyl colamino formyl methyl pyridinium chloride	—	—	—	—	—	1.5	—
Triethanolamine	0.75	0.5	—	—	—	—	—
Water	93.65	86.6	77.9	89.4	84.9	88.4	84.9

Part C

Ethyl alcohol	—	5	—	—	—	—	—
Colour	q.s	q.s	q.s	q.s	q.s	q.s	q.s
Perfume	q.s	q.s	q.s	q.s	q.s	q.s	q.s

Creams and Lotions

Emollients are materials used for the prevention or relief of dryness as well as for the protection of the skin from a biochemical viewpoint.

Formulas for emollient creams, non-ionic o/w type

Part A

Arlacel 60	3.0%	—
Tween 60	4.0	—
Atlas G-1702	—	5.0%
Atlas G- 1726 .	—	5.0

Bees wax	5.0	5.0
Lanolin	—	3.0
Hydrogenated vegetable oil	17.5	25.0
Light mineral oil	26.0	20.0
Propyl paraben	0.15	0.15
Antioxidant	0.05	0.05

Part B

Methyl paraben	0.15	0.15
Sorbitol	5.0	—
Citric acid	0.1	—
Water	38.8	36.4
Perfume	0.25	0.25

Part A

Bees wax	3.0%	—
Spermaceti	3.0	—
Light mineral oil	30.0	—
Glyceryl monostearate pure	12.0	4.5%
Lanolin	—	1.0
Isopropyl myristate	—	4.3
Polyethylene glycol 1000 monostearate	—	6.0
Stearic acid	—	7.2
Propyl paraben	0.15	0.15

Part B

Methyl paraben	0.15	0.15
Glycerol	8.0	—
Propylene glycol	—	2.5
Water	43.4	74.0
Perfume	0.3	0.2

Part A

Lanolin	1.0%	1.0%
Arlacel 83	2.0	—
Paraffin	10.0	—
Light mineral oil	15.0	15.0
Petrolatum	35.0	—
Atlas G-1425	4.0	5.0
Atlas G-1441	—	1.0
Bees wax	—	2.0
Stearic acid	—	15.0
Propyl paraben	0.15	0.15

Part B

Methyl paraben	0.15	0.15
Sorbitol	2.5	10.0
Water	29.9	50.3
Perfume	0.3	0.4

Formulas for emollient creams, nonionic w/o type

Part A

Glyceryl monostearate pure	5.0%	—	10.0%
Petrolatum	5.0	—	10.0
Light mineral oil	5.0	25.0%	10.0
Micro crystalline wax M.P 79.5°C	10.0	5.0	—
Amerchol L-101	15.0	10.0	—
Lanolin	—	10.0	—
Lanolin absorption base	—	—	15.0
Bees wax	—	—	15.0
Propyl paraben	0.15	0.15	0.15

Part B

Methyl paraben	0.15	0.15	0.15
Water	59.4	49.4	39.4
Perfume	0.3	0.3	0.3

Part A

Peanut oil	8.0%	—	—
Protegin X	26.0	—	—
Spermaceti	5.0	5.0%	—
Lanolin	5.0	5.0	—
Cetyl alcohol	2.0	2.0	1.0%
Glyceryl monostearate pure	1.0	10.0	—
Light mineral oil	5.0	30.0	23.0
Alcotec DS	—	2.0	—
Bees wax	—	—	7.0
Alcolan absorption base	—	—	27.0
Propyl paraben	0.15	0.15	0.15
Antioxidant	0.15	—	—

Part B

Methyl paraben	0.15	0.15	0.15
Glycerol	3.0	—	—
Water	44.2	45.4	41.4
Perfume	0.35	0.3	0.3

Part A

	I	II
Peanut oil	10.0%	—
Petrolatum	3.0	—
Spermaceti	4.0	5.0%
Bees wax	12.0	12.0
Light mineral oil	23.0	27.0
Lanogene	—	6.0

Propyl paraben	0.15	0.15
Antioxidant	0.15	—

Part B

Methyl paraben	0.15	0.15
Borax	0.7	0.7
Water	46.6	48.7
Perfume	0.25	0.3

All Purpose Creams

These are generally based on a typical cold cream formula. They contain stearic acid, humectant, and lanolin. By replacing part of the oil phase with stearic acid oiliness of the cream is reduced. Lanolin and humectant are emollients. Since they are not so oily they can be used in most cases where a vanishing cream is normally used.

Formula for all purpose cream o/w nonionic type

Part A

	I	II	III
Tween 85	1.0%	—	—
Arlacel 85	1.0	—	—
Tween 40	—	8.0%	—
Atlas G 1705	—	2.0	—
Atlas G 1425	—	—	5.0%
Atlas G 1441	—	—	1.0
Bees wax	2.0	2.0	2.0
Lanolin	4.0	4.0	1.0
Stearic acid	15.0	10.0	15.0
Light mineral oil	23.0	20.0	23.0
Propyl paraben	0.15	0.15	0.15

Part B

Methyl paraben	0.15	0.15	0.15
Sorbitol	12.2	12.0	10.0
Water	41.2	41.4	42.4
Perfume	0.3	0.3	0.3

Formula 1 is a thixotropic type of cream. Emollient ingredients like lanolin, or its derivatives, cetyl alcohol, spermaceti, and cocoa butter can be added to leave an emollient oily film on the skin.

Therapeutic Creams

The set of formulations discussed here may be used as basic preparations on which one may choose to add active therapeutic agents. The chemical nature of the particular ingredients must be considered in relation to its compatibility with certain emulsifier types. A water-soluble antihistamine hydrochloride for example is cationic in behavior and would therefore be incompatible with anionic emulsifiers. Either o/w or w/o emulsion may be used, depending upon the temperature effects preferred on the skin.

Once the cream has been brought down to filling temperature, the problem of transfer to the filling hopper arises. The cream is either pumped or moved by gravity. The latter is preferred because any shear action due to pumping can be avoided lest it has an adverse effect on the consistency of the cream. The type of filling unit used is also important. Piston filler is preferred over the gear type because the former exerts less of a breaking action on the structure of the creams.

Problems do not just end with manufacturing. Storage too presents its share of problems. Hand creams stored in a warm place could have the following effects.

1. White creams tend to turn off-white in colour (towards yellow).
2. Stearate creams are likely to develop a pearly shean, which may not be desirable.

3. The consistency may become soft; it depends of course on the temperature and time of exposure. This may not be reversible.

Storage in cold places has its problem as well. At temperature ranging from 5–15°C syneresis occurs. In other words, with the shrinking of the gel structure and forcing out of the aqueous phase which appears in the form of droplets on the surface of the cream and as a continuous film on the inside walls of the jar. The phenomenon is common in colloidal systems and in hand creams most common in the soap type. Nonionic emulsifiers as adjuvants will help in reducing such an occurrence.

Hand lotions are quite similar to hand creams, only differing substantially in the total solids proportion.

Acidic Cleansing Cream

Normal human beauty skin is covered with an acid layer pH 3–5. But pH 5–6 is more desirable to protect the skin from bacterial infections. But a sick person may not have this pH and also those who use excessive soap. In the alkaline pH, the skin is more susceptible for attack by pathogens.

Hence, acidic cleaning creams appeared in the market.

A typical example of such a cream is

Wool wax	2.0%
Anhydrous lanolin	9.0
Stearyl alcohol	2.0
Petrolatum	35.0
Bees wax	2
Glycerol	4
Lactic acid	1.5
Water	44.5

Procedure

All the waxy materials are melted at 55°C. Aqueous phase containing water, lactic acid and glycerol is also heated to 57°C and both the phases mixed and stirred till the temperature is cooled to 45°C, when perfume is added and passed through a colloid mill and, stored for packing.

Detergent Cleansing Creams or Soap Creams

Lard	20%
Coconut oil	12
Potassium hydroxide	12
Bitter almond oil	3.5
Perfume geranium	5
Aqua	q.s

Procedure

All the waxes are melted at 40°C. The alkalis are dissolved in water and added to the molten waxes. Stirring is continued till solidified mass is obtained

Polyethylene glycol 400 monosterate	10.0%
Stearic acid	16.0
Potrassium hydroxide	0.8
Glycerol	5.0
Water	67.5
Preservative	0.2
Perfume	0.5

Procedure

Heat the aqueous phase with allied ingredients and add to the molten waxes phase and continue stirring till the mass solidifies.

Antibacterial Creams

After the natural and synthetic, soap creams, antibacterial creams appeared in the market. With hexachlorophene as the antibacterial compound mostly.

Formula

Beeswax	9%
Paraffin	10
Mineral oil 65 ps	30
Cetyl alcohol	1
Deltyl extra	10
Hexachlorophene	0.5
Borax	1
Water	38
Perfume	0.5

Procedure

Dissolve the hexachlorophene in the deltyl extra. All the waxy materials are melted in a bath at 70°C. Borax and water are also heated and mixed with molten wax phase or stirred till the temperature drops to 50°C when the perfume is added and stored for packing.

Hand Lotions

The one major difference in formulating hand lotions and hand creams is the maintenance of the lotion in a stable fluid state or flow characteristic. This fluidity of hand lotions must be kept within certain acceptable limits. If the viscosity is very low, the lotion literally "runs" out of the bottle, which is out of acceptable limits. It is equally considered out of acceptable limits if the lotion exhibits gelling tendencies, which make it difficult to pour from a bottle. The problem of developing hand lotions with ideal viscosity is not an easy one. It becomes even more difficult to predict the viscosity pattern of a lotion at the time of formulation over the shelf life period (about 2 years). In fact there does not exist a method to determine the viscosity of an emulsion in advance at any given time in the future. However, it is only through experience excessive gelation or low viscosity can be prevented.

The problem of gelation occurs mainly in stearate type lotions.

Furthermore, excessive mechanical action of these type of lotions leads to low viscosity because they are sensitive to shear action. A stearate type lotion though thin in viscosity initially does not necessarily remain so. As a rule the viscosity increase with time. Of course temperature fluctuations also have a direct effect on viscosity. Shelf life studies at prevailing room temperature show that viscosity of stearate type hand lotions tends to thicken with time.

Therefore, steps need to be taken to prevent gelation during the shelf-life of the hand lotion and by observing the following precautions it could be prevented or atleast retarded.

1. Avoid excess amounts of polyol fatty esters, like glyceryl monostearate, or fatty alcohols, such as cetyl alcohol. The quantity considered excessive varies with each system; generally for a stearate type lotion 0.5% is just about right. In this type of hand lotions containing ethyl alcohol, 1% cetyl alcohol can be in excess.
2. The dispersed wax phase can be plastisized with high concentrations (10%) of mineral oil.
3. Small amounts of an alkyl sulphate (0.1–0.5%) such as sodium lauryl sulphate can be incorporated into the formula.

The nonionic and cationic types of hand lotions too show tendencies to gel. Proper balancing of emulsifiers, and lipophilic fatty acid esters and fatty alcohols in the formulation usually prevents gelation.

Hand lotions with low initial viscosity for a period of say about 4 weeks can be problematic because they may be too thin to accept when they reach the consumer. This is attributed to the thixotropic nature of the stearate gels. The term thixotropic has created some misconceptions in connection with emulsions. Originally it was used to describe a reversible isothermal gelsol gel transformation; or in other words it applied to those gels that break down on being shaken and reset on standing. In the present context however, it refers to the property of reversible alteration in flow characteristics when work is performed on them, the alteration here being: the greater the fluidity with increased agitation or shearing action. Armed with this understanding of thixotropy and knowledge of stearate gels, it is clear that the low initial viscosity is directly related to time, rate, and type of stirring.

In practice to reduce the time of shearing action on the lotion, the following steps may be taken.

Cool rapidly to a temperature 5-10°C below the gel point of the system.

The manufacturing procedure for hand lotions and creams basically involves the same operations. The oil phase is heated and added to the water phase. The order, rate and temperature of the addition are important for the stability of the emulsion, hence it must be established and maintained. A light stirrer is used for mixing in hand lotions, and when sufficient emulsifying agents are used, this agitation by a propellar yields a finer particle size than homogenization or milling. A rheostat is used to control the speed of stirring to avoid a vortex and the possibility of entrapping air.

As we have already seen the rate of cooling has to be very efficient in order to avoid excessive stirring, which leads to thinning of the emulsion. Perfume and alcohol should be incorporated at a temperature below 50°C.

The filling temperature of hand lotions differs from system to system depending on their viscosity characteristics. Usually it is about 25–30°C. However, the lotion is left to stand overnight at room temperature to allow entrapped air to surface. The filling is done by gravity and a mild vacuum fill may be employed to avoid formation of foam.

Recommended Equipment

Stainless steel equipment should be used in the manufacture of hand creams and lotions, glass lined tanks can be used but use of copper, tin, and iron should be avoided.

CONCLUSION

For a long time chemists have been haunted by two questions

1. How can the physical gelation tendencies of a hand lotion be evaluated so that one can predict whether or not the product will gel during its shelf life?
2. How can physical properties of a hand cream be evaluated so that one can predict its consistency pattern over its normal shelf life?

The answer to the first question has been sought in the colloidal phenomenon known as rheopexy (causing a thixotropic system to gel by regular movements, such as tapping). So far, the efforts have proven unsuccessful. The answer to the second has eluded even the outstanding colloid and physical chemists.

The ageing hands pose a natural problem and more attention. As far as progress in hand treatment is concerned, the search for a universal protective hand product should be encouraged. Presently very little has been done in this field.

Perhaps for a complete answer, the study of the basic physiological and chemical changes which occur in the skin during ageing need to be followed.

3

Shaving Cream

Shaving Soaps and Creams

S HAVING SOAPS WERE first made in USA about 125 years ago. It was an improvised home made potash soap which was prepared by saponification of kitchen fats with crude potash made from wood ash and quick lime. Caustic soda was found to be a necessity to obtain a solid cake soap.

Theoretically, solid shaving soaps and lather creams are somewhat similar products. They differ only in physical appearance and forms. The brushless shaving cream however differs considerably from the above two solid preparations.

A shaving cream, in any form should conform to certain specifications in order to make shaving a comfort.

1. The shaving cream must be nonirritating to the face and must retain its moisture as long as it is on the face.
2. It must soften the beard sufficiently so that the razor cuts readily.
3. The shaving preparations must provide lubricity so that the razor glides smoothly on the face and should be of sufficient viscosity to hold each hair erect.
4. Further, the product should remain stable over a wide temperature range.

5. It should be noncorrosive and non-rusting and should be able to be washed down the drain without clogging.

6. If it is a brush shaving cream, it must be able to develop a high lather very rapidly.

Brushless Shaving Cream

The function of soap and lather cream as shaving aids is by rapidly softening the beard. However, it is not true of brushless shaving cream which acts more slowly. In other words it has relatively lesser and considerably slower beard softening qualities. The primary objective of a brushless shaving cream is to keep the whiskers in a moist condition during the shaving and provide lubrication for. Most of the beard softening effect is actually obtained by preshave treatment with soap and water. Hence certain adjustments are needed in the formulation. For example brushless shaving creams need to contain wetting agents rather than lathering agents. On the other hand it is necessary for the lather cream to have a higher pH (about 10) than the brushless (about 8) because a higher pH is conducive to more effective hair softening.

Solid Shaving Preparations

For the manufacture of bar shaving soap, potassium hydroxide must be of highest purity relatively free from iron and sulphates and low in potassium chloride. In the manufacture of any solid shaving soap perfume is incorporated during the milling process. In order to increase whiteness some titanium dioxide is suggested of the order of about 0.2%.

Formula 1 Bar Cake Shaving Soap

Toilet soap form kettle	50.00%
Coconut Oil, Manila	5.00
Stearic acid	10.00
Tallow prime	20.00
Caustic soda, Sp. Gr. 1.5	2.30

Caustic potash, Sp. Gr. 1.5	10.00
Antioxidant	0.05
Water	2.65

The solid shaving soap may be manufactured in several ways.

1. To the mixture of tallow and coconut oil soaps, a certain amount of potash is added. It can entirely be made in a soap crutcher using a mixture of sodium and potassium hydroxide. This method leaves all the glycerol from the fats used in the finished cake, resulting in a slimy or sticky feel. Glycerol (upto about 10%) improves the lather and has an emollient feel on the face.
2. To add to kettle made boiled soda soap potassium stearate in the crutcher. This method leaves the soap free of glycerol, which is rather brittle and prone to crumble.
3. To saponify the requisite amount of tallow with potassium hydroxide in the crutcher, add kettle made sodium soap to it and neutralize the excess alkali with stearic acid.

Lather Shaving Cream

From a manufacturer's point of view the lather shaving cream is one of the most difficult toilet articles to make, considering particularly its stability in hot and cold weather conditions. Even changes in source of raw materials that are analytically identical in the laboratory, but differ in some slight respect, may cause difficulties in manufacture or storage. Essentially a lather cream may be considered similar to a shaving bar soap except for the consistency. The lather cream apart from presenting similar problems of a bar shaving soap has one additional one that of viscosity stability. In short, a lather shaving cream may be described as a solution of soap in glycerol and water, in which excess soap may be dispersed. The content of potassium chloride must be kept to a minimum because of its ionic effect on the physical properties. Borax is another valuable ingredient but it must be carefully formulated for the same reason as potassium chloride. The most desirable shaving cream is one which does not change its viscosity even in wide range of temperatures. It is particularly for this reason the salts and the other ingredients have to be very carefully selected.

Choice of Electrolytes

The choice of electrolytes like chlorides is a tricky one because of their corrosive effect on the metal tube. However, they can be countered by corrosion inhibitors. Extensive shelf life tests are needed over a wide range of temperatures while selecting the electrolyte and the metal packaging.

Stearic Acid

Yet another very important ingredient for lather cream is stearic acid. There are 3 grades of this material commonly sold; single-pressed, double-pressed and triple-pressed. Normally a double-pressed stearic acid is preferred in order to obtain a white lather cream. A qualitative test to find out whether the stearic acid is suitable is to add an excess of strong potassium hydroxide solution to a small sample of melted stearic acid. The soap formed should not be brown or yellow. The odour of stearic acid is another important aspect. A poor grade stearic acid easily develops rancidity and could spoil the entire batch.

Triethanolamine

Triethanolamine should be avoided in lather cream, because it often causes discolouration upon ageing. Further, there are no advantages to be gained from its use.

Coconut Oil

A fraction of the acids derived from coconut oil is necessary in a shaving cream, the sodium or potassium soaps made from coconut oils are very soluble in water and lather very freely. The lather is large bubbled, thin and breaks down readily. Use of specially deodorized oil is recommended in order to avoid odour.

Oleates

Oleates are generally present in lather creams as well as shaving soaps. Coconut oil itself will provide sufficient oleates for most formulations. However, the oleates have a desirable and undesirable property for a shaving cream. They are free-lathering, non-irritating, and tend to keep the lather moist. But if in excess, they will make the cream stringy and have a tendency to go rancid.

Fatty Acids

Fatty acids which are relatively more unsaturated than oleic acid should not be present in significant amounts because their soaps tend to become rancid.

Lather Characteristics

Acid	Lather from Sodium soap	Lather from Potassium soap
Stearic	Almost non-existent; soap very soluble	Slimy dense, not copious
Palmitic	Poor	Dense but not copious
Myristic	Not copious	Very good, copious
Lauric	Large bubbles, copious	Very light, unstable
Oleic	Readily formed, stringy	Very readily formed not stable

Manufactures over the years have been striving to improve the lather of shaving creams, softening (of the beard) quality and moisture retention to make the face feel better.

Mostly superfatting agents have been used because of their two-fold purpose. One to neutralize excess alkali and second to stabilize the foam by retaining the moisture. For the last mentioned purpose, glycerol, lanolin, and its derivatives, and carbowaxes are very useful. Lanolin is widely used in modern shaving creams as a superfatting agent because apart from being a superfatting agent it is an effective emollient. But lanolin and its derivatives, used in quantities in excess of 1-2%, will interfere in lathering properties and create difficulties in perfuming the cream. Other superfatting agents are, stearic acid, free fat from unsaponified coconut oil, mineral oil (rarely used) and vegetable oils (tend to become rancid).

Substances known to reduce surface tension and produce a fine-bubble lather are saponins, sodium chlorate, and lecithin. Small quantities of antioxidants like phenolic compound with long side chains, are used for the preservation of the cream.

Use of fractionated acids from coconut or palm kernel oils renders shaving soaps nonirritating and enhances lathering.

Humectants

Glycerol has a very desirable property of making a cream soft and maintaining it as well thus allowing for easy extrusion from a tube. In reasonable amounts, it improves the lather and has an emollient feel on the face. Some creams have used as much as 15% in their formulations. Sorbitol was suggested as an alternative to glycerol but was not found to be as effective.

Menthol

Menthol is used to give a cooling effect. But it has not received universal acceptance.

Formulas for Lather Cream

	2	3	4	5
Stearic acid	35.0%	21.0%	20.0%	38.8%
Coconut oil	10.0	10.5	6.0	9.7
Potassium hydroxide	6.1	6.6	7.5	8.0
Sodium hydroxide	2.0	0.4	0.5	1.6
Glycerol	8.0	7.0	12.0	11.6
Water	37.4	53.2	37.5	30.3
Additive	1.5	1.3	1.0	—
Boric acid	—	—	0.5	—
Stearine	—	—	15.0	—
Preservative	q.s	q.s	q.s	q.s
Perfume	q.s	q.s	q.s	q.s

Procedure

After saponification of stearic acid and coconut oil, add the remainder of stearic acid and also add special ingredients such as lanolin and the anti-oxidant. Heat the glycerol, boric acid (if required) and half of the water required to 65°C, mix and run slowly in the crutcher stirring till the creamy paste is uniform. Heat the remaining water to 44°C and add as quickly as the cream absorbs it.

BRUSHLESS SHAVING CREAM

Lather and brushless shaving creams differ in purpose as well as composition. A leather shaving cream is a cream like soap, a brushless product is not a soap in the true industrial or chemical sense of the word. It is rather a vanishing cream with an added lubricant. A brushless shaving cream is an oil-in-water emulsion relatively easy to make and relatively stable.

While the lather cream is designed to soften the beard, the brushless cream requires pre-shave treatment like washing the beard with (hot) water and soap. The brushless shaving cream sustains the softening of the beard by the preshave treatment throughout the shave.

The brushless shaving cream exceeds the list of ingredients suggested for lather shaving cream. Apart from the common ingredients, like stearic acid, coconut oil or other oils containing fatty acids and lanolin, it includes the gums, such as karaya and Irish moss. Also mucilaginous material such as methylcellulose, polyvinyl pyrrolidone, sodium carboxymethyl cellulose have been suggested. They not only provide body to the cream but are good stiffening and water retaining agents.

Brushless shaving creams are formulated around the following general formula.

Stearic acid	10–20%
Preservaticve	0.2
Mineral oil/petrolatum	3–13
Base	0.5–2
Lanolin	0–5
Gums or thickeners	0–0.5
Water	60–75

Several preparations that are variations and departures from the general formula are shown in formulas 6–11.

Formulas for Brushless Shaving Cream

	6	7	8	9	10	11
Stearic acid, triple pressed	10%	15%	22%	26%	14%	15%
Cetyl alcohol or stearyl alcohol	—	3	—	—	—	—
Petrolatum	10	5	—	—	—	13
Mineral oil, heavy	—	—	3	9	8	—
Carbowax	—	5	—	—	—	—
Glyceryl monostearate	5	—	—	—	—	—
Sodium hydroxide	—	—	—	—	0.4	—
Triethanolamine	1	—	0.75	0.5	0.6	—
Methyl *p*-hydroxy benzoate	0.2	0.2	0.15	—	—	—
Boric acid	—	1.0	—	—	0.7	2
Borax	—	—	0.5	0.5	—	—
Lanolin	—	1.0	3.6	—	5	—
Water	73.8	68.3	70	64	71.3	69.5
Span 20	—	1.0	—	—	—	—
Tween 20	—	0.5	—	—	—	—
Potassium hydroxide	—	—	—	—	—	0.5
Perfume	q.s	q.s	q.s	q.s	q.s	q.s

Apparatus

Shaving soap manufacture requires the standard equipment of a general soap factory i.e., soap kettles, driers, plodders, milling machines and crutchers.

PRESHAVE PREPARATION

The primary purpose of preshave preparations is to prepare the beard and the skin of the face more effectively than the shaving preparations alone.

This is accomplished by better softening of the hair, increasing the lubricity of the shaving soap lather and reducing the sensivity of the skin from mechanical and chemical effects of shaving.

Beard Softeners

Although brushless shaving creams are high in lubricity they do not soften beard quickly, more so without the preshave treatment like washing with hot water and soap. Emulsification of the natural oil on the beard and suspension of the facial soil by a solution of soap or synthetic detergent applied before the brushless shaving cream, are highly effective in hastening the wetting and softening of the beard by water contained in the shaving cream.

Formula 1

Duponol WAT	20.0%
Aerosol OT - 100%	0.1
Carbitol	2.0
Ethyl alcohol specially denatured	8.0
Water	68.9

Procedure

Dissolve the Aerosol OT 100% in a mixture of Carbitol, alcohol and water. Add the Duponol WAT and mix until uniform. Other brands of triethanolamine lauryl sulfate and dioctyl sodium sulfosuccinate can be used in place of Duponal WAT and Aerosol OT 100% respectively of equivalent content of active ingredient.

Formula 2

Coconut oil fatty acids, double distilled	4.20%
Oleic acid, low linoleic content	5.6
Propylene glycol	5.0
Triethanolamine	2.85
Monoethanolamine	1.26
Tergitol NPX	2.0
Water demineralized	79.09

Procedure

Mix the fatty acids and stir in the propylene glycol. Add the amines and stir until a clear solution emerges. No heating is required. Add the tergitol NPX and dilute with water.

Skin Conditioner

The conventional shaving soaps and the lather shave means lack in beard softening action. The formula hereunder is a brushless shave cream reformulated to increase its beard softening, moisturizing, skin-lubricating and skin protecting properties. Menthol and camphor are added for their cooling effect on the skin and a suitable soap compatible antiseptic can be incorporated into the product.

Formula 3

Part A

Stearic acid, triple pressed	20.6%
Diglycol stearate, self-emulsifying	2.5
Mineral oil 55/65	4.0
Lanolin anhydrous	1.0
Sulfonated castor oil 70%	1.0

Part B

Triethanoalmine	1.3
Borax USP	0.9
Water	64.0
Propylene glycol	4.0

Part C

Menthol	0.1
Camphor	0.1
Perfume oil	0.5

When sodium alginate and similar natural gums are prepared in advance it is advisable to preserve them with 0.1–0.2% of methyl-*p*-hydroxy-benzoate.

Procedure

Heat A until melted and homogeneous to 70°C. Prepare a 2% mucilage of sodium alginate and then add ingredients of B and heat to 70°C. Add A & B with good agitation and continue down to 45°C. Stir in camphor and menthol dissolved in perfume oil. If softer consistency is desired continue slow stirring down to below 35°C.

Pre–Electric Shave Preparation

A popular mode of shaving these days is the 'dry' electric shave. A preshave preparation has a highly specialized purpose. Either the moisture on the skin is removed or prevented from interfering with the smooth passage of the cutting head of the shaver over the beard.

There are two different solutions for this problem. One employs talc to absorb moisture and to leave its characteristic slip on the face. The second uses alcoholic lotion to dry and tauten the skin by its astringency and to preferably leave a lubricating film on the skin.

A talc stick is made for this purpose with the help of a suitable binder, calcium sulfate (plaster of paris) with the talc and other ingredients, moistened with water, and then moulded. An excellent binder for powdered sticks is colloidal magnesium aluminum silicate (veegum). The addition of a metallic soap, such as zinc stearate, increases slip and improves adhesion. Emollients may be added with careful consideration such that the 'slip' or interference with the stick to dry the skin is not reduced.

The formula hereunder is as illustration of a talc stick that is suitable for a variety of electric shaves.

Formula 4

Part A

Zinc stearate	5.0 parts
Iron oxide Pigments	q.s
Light Magnesium carbonate	2.0
Perfume	q.s
Talc to make	100.0

Part B

Veegum	1.5
Water	30.5

Procedure

Adsorb the perfume completely on magnesium carbonate, add the zinc stearate and pigments and disperse all, thoroughly in the talc. Prepare an aqueous solution of veegum by adding slowly to the water with good agitation until smooth. Add B to A and mull to a smooth paste.

Another method does not require the binder in the talc. Instead pressure is applied (450-600 psi) and the stick is coated with a suitable material to protect it from cracking, crumbling and chipping. Commonly used for this purpose are vinyl resins such as polyvinyl-chloride and polyvinyl acetate dissolved in ethyl acetate.

The alcoholic preshave lotion is either astringent or oily. The first property makes the hair stiff and dry thereby making it stand upright. This is achieved by using a high concentration of alcohol for its dehydrating effect and adding a mildly astringent material, such as lactic acid or zinc phenolsulfate. Menthol and camphor serve to cushion minor trauma.

The oily type lubricates the beard and skin by depositing a thin film of oil on the face. This prevents drag and pull of the cutting head against the skin, especially in warm, humid weather and generally improves performance by reducing friction.

Formula 5 (Astringent type)

Zinc phenolsulfate	1.8%
Ethyl alcohol, specially denatured	40.0
Menthol	0.1
Camphor	0.1
Distilled witch hazel extract	58.0

Procedure

Dissolve the zinc phenolsulfonate, menthol, and camphor in the alcohol and dilute with the distilled witch hazel extract, colour and filter clear.

Formulation 6 (Lubrication Lotion Oily)

Isopropyl myristate	20.0%
Ethyl alcohol, specially denatured	80.0
Perfume oil	q.s

Procedure

Dissolve the isopropyl myristate and perfume oil in the alcohol add colour and filter clean.

AFTER SHAVE PREPARATION

The general function of an after shave preparation is to relieve discomfort and irritation caused by shaving. Its purpose is to cool and soothe the skin, impart a feeling of freshness and well being. The alcoholic lotion is the popularly used after shave preparation. Here are some after-shave preparations.

Formula 1

Bay oil	0.20%
Pimenta oil	0.05
Ethyl alcohol	50.00
Jamaica rum	10.00
Water	39.75
Caramel	q.s

Formula 2

Myristica oil	0.08%
Orange oil	0.05
Pimenta oil	0.05
Ethyl alcohol	61.00
Water to make	100.00

Procedure

Mix the oils with alcohol and gradually add water until the product measures 100 ml. Set the mixture aside in a well closed container for eight days, then filter using 10 gm of talc, to render the product clear.

Formula 3

Peppermint oil	1%
Glycerol	5
Bay rum	94

Formula 4

Potash alum	2%
Glycerol	3
Menthol 1% in ethyl alcohol	5
Orange flower water	20
Rose water	20
Witch hazel extracts	50

The aromatic waters mentioned are saturated solutions of odoriferous principles prepared by distilling the respective plant materials with water, separating the excess volatile oil, if any, from the clear water portion of the distillate and adding alcohol if necessary for preservation.

Clear Lotions

Hereunder are some characteristics, atleast to some degree, of an after shave, which manufactures endeavour to incorporate in their product.

1. Relief or irritation and tension of the freshly shaven skin.
2. Cooling and refreshing action.
3. Mild astringency.

4. Neutralization of some soap left on the skin to help restore neutrality.
5. Antibacterial activity/action.
6. A pleasant, long-lasting and characteristic fragrance.

AFTER SHAVE PREPARATION

Concentrations of above 60% by volume of alcohol in an after shave preparation caused excessive sting and smarting. The ideal concentration varies from 40–60 % by volume.

Concentrations of menthol vary from 0.005% to 0.2%. The ideal concentration of menthol is 0.1%. It then cools the facial skin after a close shave. Higher concentration of menthol causes an undesired effect or feeling of burning sensation.

After a shave with soap the facial skin is alkaline. Small quantities of week acids like boric acid, lactic or benzoic acid in an after shave lotion help restore the normal slightly acidic condition of the skin.

The addition of aluminum or zinc salts increases astringency and styptic action. The aluminum or zinc sulphates are insoluable in alcohol and are hence unsuitable for lotions with light alcoholic content. Anyway, the styptic action of arresting bleeding from minor razor cuts is difficult to achieve because of the solvent action of alcohol on the blood clot.

Emolliency is imparted through polyols, like glycerol, propylene glycol and sorbitol by incorporating them in low concentration of upto 3%. All these polyols are considered to be innocuous. Propylene glycol is the most preferred because of its low viscosity and high volatility. Further, the polyols do not leave the skin greasy or tacky.

Lipophilic emollients such as lanolin and its derivatives, hydrocarbons, phospholipids and fatty acid alcohols and esters are difficult to incorporate in conventional alcoholic after shave because of their limited solubility. Moreover they tend to reduce the characteristic cooling and refreshing "after feel" of these preparations. They are however employed to good effect in emulsifying creams and lotions.

The use of bland antiseptic in shaving preparations has been recommended as a prophylactic measure.

The formula here under is a simple, modern aftershave lotion illustrating the use of mild acid to correct the alkaline reaction of the skin after shaving with soap.

Formula 1

Ethyl alchol	50.0%
Sorbitol	2.5
Perfume oil	0.5
Menthol	0.1
Boric acid	2.0
Water demineralized	44.9

Procedure

Dissolve all ingredients completely in the alcohol and dilute with water using good agitation. Allow to stand, preferably with adequate chilling until poorly soluble constituents of the perfume oil have agglomerated and then filter clear. If colour is to be added it is to be done at room temperature.

Formula 2

Formula 2 is an antiseptic after shave lotion containing a quaternary ammonium salt, a surface anesthetic and menthol.

Hyamine 10.X	0.25%
Ethyl alcohol specially denatured	40.0
Menthol	0.005
Ethyl p-aminobenzoate	0.025
Water	59.720
Perfume oil	q.s

It is to be noted that very low concentrations of menthol can be potentiated in action by the addition of a small amount of a surface (local) anesthetic such as benzocaine (ethyl p-aminobenzoate)

Stick Lotions

Stick preparation, which is a popular solid after shave lotion, is convenient particularly when travelling.

Formula 3

Ethyl alcohol specially denatured	80.5%
Perfume oil	1.4
Sodium stearate	6.0
Glycerol	4.0
Propylene glycol	3.0
Menthol	0.1
Water	5.0

Procedure

Place all ingredients (except perfume oil) in a stainless steel steam jacketed kettle fitted with an agitator and water cooled condenser. Heat with stirring and at 55°C add perfume oil through an addition funnel. Continue heating to reflux temperature and stir until completely dissolved. Adjust temperature to 71–74°C and pour into moulds. Colour may be added by dissolving in water of the formula.

Solid and Liquid Creams

After shave preparations can take the form also of creams both solid and liquid.

Men who find alcoholic after shave lotions uncomfortable or irritating, particularly after over exposure to sun, wind or inclement weather frequently use emollient vanishing cream or hand lotion to finish off the shave.

Formula 4

Stearic acid triple pressed	18.0%
Potassium hydroxide USP	1.2

Glycerol	5.0
Water	25.8
Distilled witch-hazel extract N.F.	50.0
Preservative	q.s

Procedure

Dissolve the potassium hydroxide in water, add glycerol and pre-servative and heat to 80°C. Melt stearic acid in a separate vessel and heat to 75°C. Add the akali solution slowly to the melted stearic acid with good agitation. When mixture cools to 50°C add the witch-hazel extract slowly with good mixing and continue slow mixing until cool. Cover and let stand overnight. Remix briefly before pack-ing.

Formula 5

Stearic acid triple pressed	15.0%
Potassium hydroxide USP	0.5
Sodium hydroxide USP	0.18
Cetyl alcohol	0.5
Isopropyl myristate	3.00
Glycerol	5.00
Water	75.82
Preservative	q.s
Perfume oil	q.s

Procedure

Dissolve potassium hydroxide and sodium hydroxide in water, add glycerol and preservative and heat to 80°C. Melt in a separate vessel stearic acid, cetyl alcohol and isopropyl myristate and heat to 75°C. Add the alkali solution slowly to the melted oily phase with good agitation. At 45°C add perfume and continue slow mixing until cool. Cover and let stand overnight. Remix briefly before pack-ing.

Formula 6

Stearic acid	3.0%
Cetyl alcohol ·	0.5
Glycerol	2.0
Methyl paraben	0.2
Quince seed mucilage	40.0
Triethanolamine	0.8
Water	48.0
Ethyl alcohol specially denatured	5.0
Perfume oil	q.s

Procedure

Heat together the glycerol, methyl paraben, quince seed mucilage, triethanolamine, and water to 75°C. Heat separately stearic acid and cetyl alcohol to 75°C. Add the two phases with good agitation. When cooled to 40°C mix alcohol, perfume and add slowly to the emulsion with stirring. At 30°C discontinue stirring and bottle. The formulas here under illustrate respectively a soft cream and a heavy lotion.

Formulas 7 and 8

Glyceryl monostearate, pure	10.00%	3.00%
Stearyl alcohol	3.00	1.50
Sorbo	5.00	2.50
Emcol E-607.5	1.00	1.00
Emcol E-607	0.25	0.25
Sodium benzoate	0.10	0.10
Perfume oil	0.30	0.30
Water	80.35	91.35

Procedure

Heat glyceryl monostearate and stearyl alcohol together to 70°C. In another vessel, dissolve the sorbo, Emcols and sodium benzoate in water and heat to 70°C. Add the oily phase to aqueous phase with good agitation and continue mixing while cooling. Add perfume at 40°C. Mix until cooled to 25°C and package.

Powders

Aftershave powders are traditional and still are popular accessories to the shave.

Their obvious function is to impart matte finish to the face, making the shine left by a too-oily brushless shave or toning down the "too scrubbed" look after a lather shave. The masking function covers minor skin defects and hides an inadequate shave. In dark haired and heavily bearded individuals powder is indispensable in rendering a clean shaven appearance. The less obvious but important functions are identical with those mentioned for after shave lotions.

The formulation of an after shave powder differs from that of a face powder in several ways.

1. Firstly, in an after shave powder, the slip and adherence are relatively more important than covering powder and bloom. Because the slip enhances the cooling and refreshing effect.

2. It must not be opaque so that its appearance on the face is less obvious.

3. The aftershave powder should be able to absorb moisture from the skin without caking or streaking because the face usually is difficult to dry completely after the shave. However, its oil absorbing functions may not be of the same degree as that of a face powder.

4. An after shave contains a relatively large proportion of talc.

5. Colour of an after shave powder is normally light or almost suntanned flesh tone.

Formula 9

Formula 9 is a typical example of an after-shave powder.

Talc	80%
Kaolin colloidal	10
Zinc stearate	5
Precipitated chalk	3
Boric acid	2
Yellow ochure	q.s
Perfume oil	q.s

The commercial manufacture of after shave powders does not differ materially from that of a face powder.

Styptics

Styptic pencils and alum blocks are shaving accessories. The styptic pencils are exclusively to stanch bleeding from minor cuts produced during shaving. Alum blocks offer a convenient method of applying a high concentration aluminum salt to a limited area of the face.

4

Lotions

LOTIONS CAN BE defined as medicated washes. But in cosmetic science language, a lotion is a liquid preparation, applied externally on the skin to produce or enhance a beautification. The main functions of lotions are soothing and emolliency, extended to astringency, skin freshening, bleaching and other medicinal properties.

Some of the liquid creams can also be designated as lotions, although these do not contain any germicides. A general formula for a lotion includes, alcohol, water and glycerin, with some special astringents, gums, honey or antiseptic etc.

Generally denatured alcohol is used for reducing the cost of the product. Rarely isopropyl alcohol can be used as a substitute for ethyl alcohol, as methyl alcohol is not advisable because of its toxicity even through percutaneous absorption. The most commonly used gums are karaya, acacia, gum tragacanth which is a dried mucilaginous exudation, probably through insect action. But this gum gives a thicker mucilage than karaya. Gum arabic (acacia) is a true gum and is freely soluble in water. But this is mainly used as an emulsifying agent. Recently tragum has entered into market, a 2% solution can give a good stable suspension of solids.

Gum benzoin is not a true gum but resin. It has a pleasing odour like vanilla, good preservative properties as it contains, benzoic and cinnamic acids. Quince seed aqueous extract is another good suspending agent but expensive.

Marsh mallow root starch is also used. Sometimes a 1% solution of corn starch is another cheaper substitute for low expensive preparations as it gives a good mucilage. All mucilages have disadvantage of changing their viscosity on ageing and thereby causing shelf life problems.

An astringent is a substance, which contracts tissue and thereby lessens secretions. Protein matter is coagulated. Eg. Aluminum salts, zinc salt, tannins, witch hazel are generally selected for lotions normally at dosage levels of 1%. Tannic acid or tannin is a yellowish, glistening, scaly powder, soluble in 5 parts of water and in two parts of alcohol. It is incompatible with proteins, chlorides, heavy metal salts, starch, oxidizing substances and alkaloids. It is used often in external preparations as styptic antipruritic, for the treatment of burns and for certain skin disorders. It hardens the skin, when used repeatedly in concentrated solutions. Witch hazel has a soothing and astringent action, it has extensive use in lotions. It is a saturated aqueous extract of the dried leaves of Hamamellis virginiana, which are collected in autumn. The official (NF) extract contains besides water, 14% of alcohol, tannin, 8% gallic acid, volatile oil, and a bitter principle. It has a pleasant odour with good astringent action, soothing properties, and its miscibility with alcohol, glycerin and water in all proportions makes witch hazel a desirable ingredient for lotions.

Another category of desired ingredients in lotions are antiseptics and bacteriostatics. Cooling effect is also desirable for lotions. Menthol is generally used for this purpose. Lotions do need preservatives to prevent bacterial contaminations, and hence, salicylates, benzoates and parabens are used in the formulation.

Lotions are filtered through simple filters, and filteraids.

Finally raw materials used should be free from impurities hence observation is recommended for sedimentation for 24 hours before filtration. Incompatibilities should also be borne in mind before designing a formulation.

HAND LOTIONS

These are meant for keeping the hands soft and soothing, as the hands are most used in rough and tough work like dish washing, and other household chores. Solid and liquid creams, emollient lotions that are nongreasy are also recommended. A general formula of a lotion is described below.

Formula 1

Stearic acid	3.15%
Glycerin	6.0
Potassium hydroxide	0.15
Water	79.3
Alcohol	8.5
Perfume	0.5
Quince seed (extract)	2.15
Preservative	0.15

Procedure

Dissolve the potassium hydroxide in one third of the water and add the glycerin. Bring the rest of the water to a temperature of 80°C. Add the quince seed and soak for six hours and strain through muslin. Melt the stearic acid, dissolve the perfume in the alcohol. Add to the hot potassium hydroxide, the melted stearic acid and boil for a minute. Allow the temperature to drop to 70°C and then stir it into quince seed mucilage. Stir occasionally until cool, then slowly add the alcohol, preservative and perfume.

Formula 2

Gum tragacanth	4
Tincture benzoin (Sumatra)	26
Glycerin	5
Boric acid	1
Alcohol S.D. 39 B	35
Water distilled	qs 100
Perfume	qs

Procedure

Make a mucilage of the gum tragacenth with water (2/3). Make up the tincture of benzoin. Both mucilage and tincture must be strained. Dissolve the boric acid in the rest of the water (hot) in the mixing tank. Add the mucilage, then the alcohol, in which the perfume has been dissolved. Add the tincture with stirring. Mix well strain again and fill when cold.

Formula 3

Stearic acid	1.5%
Powdered soap (Neutral or white)	1.0
Alcohol S.D. 39 B	4.0
Glycerin	5.9
Borax	2.5
Gum karaya	1.5
Water	82.95
Perfume	0.5
Preservative	0.15

Procedure

Mix the gum karaya, with alcohol and stir into half the water (warm), when dissolved strain through muslin. Heat the rest of the water, and dissolve in it the borax; add the soap, glycerin and melted stearic acid, and agitate until cool. Then add the mucilage, preservative and the perfume.

PINEAPPLE JUICE HAND LOTION

Formula 4

Irish moss mucilage 3%	37.0%
Glycerin	10.0
Alcohol	15.0
Boric acid	0.5
Tincture of benzoin	0.5

Pineapple juice	36.8
Methyl paraben	0.2

The irish moss mucilage is made by soaking 3% of cleaned irish moss in 97% by weight of water. After soaking overnight, drain off the mucilage and add the required quantity to the above formula. The procedure is similar to that of the previous formula. Upto 5% powdered soap can also be added.

MENTHOL LOTION

Formula 5

Menthol	0.2%
Powdered tragacanth	0.5
Alcohol	9.0
Glycerin	4.5
Water	85.8

Dissolve the menthol in alcohol and add to the tragacanth. Add water and glycerin and mix until a smooth mixture results.

WITCH HAZEL LOTION

Formula 6

Quince seed	2.0%
Hot water	8.0
Glycerin	16.0
Witch hazel	71.5
Boric acid	2.0
Perfume	0.5

Dissolve the boric acid in water and mix with the other ingredients.

WITCH HAZEL AND BENZOIN LOTION

Formula 7

Gum tragacanth	0.75%
Glycerin	6.25
Witch hazel	6.25
Tincture of benzoin	0.4
Liquefied phenol	0.75
Oil of rose	0.1
Alcohol	6.25
Water	79.25

LEMON LOTION

Formula 8

Pectin	2.5%
Lemon juice	9.5
Citric acid	3.0
Benzoic acid	0.15
Glycerin	5.0
Alcohol	15.0
Water	64.35
Perfume	0.5

Dissolve the citric acid in water. Add the benzoic acid and lemon juice and pectin. Then add glycerin and the perfume dissolved in alcohol.

LEMON LOTION

Formula 9

Gum tragacanth	5
Lemon juice	25
Boric acid	1.25

Glycerin	1.0
Alcohol	35.0
Colour	qs
Distilled water	qs 100

HAND LOTION

Formula 10

Gum tragacanth	5.0
Boiling water	qs to 100
Ethylene glycol or glycerin	10.0
Alcohol	12.0
Phenol	0.25
Camphor	0.25

Soften the gum in water and smoothen it by forcing through cheese cloth. Add balance of ingredients and mix well.

SKIN TONING LOTION AND FRESHENERS

These are limpid liquids with weak astringent, invigorating, stimulating and at times antiseptic properties. They are generally meant for freshening the skin and to remove any residual creams.

The technology, and skill in making these solutions adds to the elegance and quality of the product. Clarity of the product can be achieved by filtering the liquid with talc or magnesium oxide and ageing helps in mellowing the finished product, hence it is desirable.

Alcohol	30.0%
Glycerin	5.0
Lactic acid 85%	2.0
Water	62.5
Perfume	0.5

Mix perfume with alcohol and rest of ingredients, age and filter.

Formula 11

Boric acid	2.0%
Alum	1.0
Camphor	0.02
Menthol	0.14
Formaldehyde	0.1
Ethylene glycol	4.0
Water	82.49
Alcohol	10.0
Perfume	0.25

Mix the camphor and menthol together to form a liquid. Heat one part of water; dissolve the boric acid in it. Similarly with alum. Put the remainder of water into the tank, add the formaldehyde, mixed with the glycol then add the boric acid solution and mix again. Then add camphor menthol solution, alcohol and perfume. Mix thoroughly, age and filter.

Formula 12

Aromatic spirit of ammonia	2.0%
Alcohol	4.0
Lavender oil	0.25
Distilled water	93.75

Mix the aromatic spirit of ammonia with the water. Add the lavender oil dissolved in alcohol, age and filter.

Boric acid	1.0%
Witch hazel	15.0
Rose water	15.0
Alcohol	10.0
Orange flower water	59.0

Warm the witch hazel, and dissolve the boric acid in it. Mix the rest of the ingredients with the orange flower water and add the boric acid solution. Mix, age and filter as usual.

Formula 13

Tincture of benzoin	1.25%
Boric acid	0.5
Glycerin	8.3
Perfume	0.2
Alcohol	40.0
Witch hazel	9.75
Orange flower water	40.0
Colour to tint	q.s

Dissolve the tincture of benzoin and perfume in the alcohol. Warm the witch hazel and dissolve the boric acid in it. Mix the witch hazel and the orange flower water. Add glycerin then add the tincture and alcohol. Mix thoroughly, age and filter as before.

Formula 14

Menthol	0.05%
Glycerin	5.0
Alcohol	5.0
Boric acid	2.0
Bay rum	15.0
Water	72.7
Perfume	0.25

Procedure

Dissolve the menthol in alcohol and the boric acid in a small quantity of warm water. Mix the rest of the water, with the Bay rum. Add the menthol solution, mix and then add boric acid solution and the glycerin. Finally add the perfume mixed with a little precipitated chalk. Mix, age and filter as before. Add perfume to alcohol.

ASTRINGENT LOTIONS

These are intended to correct excessive oiliness and also to make coarse pores less noticeable. Oiliness is caused by either fatty diet or individual metabolic changes of fat, or by constipation. Treatment with astringent lotion in the morning is helpful in removing the oiliness, preferably in addition to a treatment with astringent cream at night.

The astringents generally react with protein. The best known astringent is tannic acid. It is even helpful in the treatment of burns because of its reaction with proteins or the decomposed protein matter to form insoluble substances. Further it prevents the absorption of the toxic products of protein decomposition and the formation of the leather like crust helps to relieve the pain and to protect the underlying sensitive surfaces. So do tannins obtained from natural resources like woods, barks, and other plant materials.

Testing of Astringent Action

1. By precipitation of protein, quantitatively this can be measured.
2. The difference in the extensibility of tissues before and after treatment with a solution of the substance.
3. The increased resistance of red corpuscles to heamolysis.

Formula 15

Alum	0.75%
Zinc sulfate	0.1
Glycerin	10.0
Alcohol	10.0
Water	78.65
Perfume	0.5

Procedure

Dissolve the alum in one part of water and the zinc sulfate with the remainder of the water. Add the alum solution and then the alcohol. Allow to stand for 24 hours and filter.

Formula 16

Boric acid	3.0%
Alum	1.3
Formaldehyde	0.2

Glycerin	5.0
Alcohol	10.0
Water	80.0
Perfume	0.5

Procedure

Dissolve the boric acid in two parts of water with heat. Similarly, alum using less water. Mix the formaldehyde with the glycerin, and the perfume with the alcohol. Then add the rest of the water the alum solution, the boric solution, the formaldehyde and the perfume solution.

Formula 17

Lactic acid	5.0%
Alum	3.0
Oxyquinoline sulfate	2.0
Glycerin	10.0
Alcohol	10.0
Water	69.5
Perfume	0.5

Procedure

Dissolve the alum in one part of water. Mix the lactic acid with the remainder of the water. Add the oxyquinoline sulfate, then the alum solution.

GLYCERIN LOTION

Formula 18

Aluminum sulphate	1.0%
Glycerin	4.5
Triethanolamine	0.5

Alcohol	30.0
Perfume	0.5
Water	63.5

Rich Massage Lotion

A heavy, rich lotion with excellent lubricity. Apply regularly. Massage the skin in circular movements.

Skin type: All types for massage, excellent for dry areas such as knees, elbows, etc.

Glycerin & Rosewater Lotion with Vitamin E

A light rose scented moisturing lotion gives the added benefits of vitamin E. Ideal for young and sensitive skin types.

Skin Type: Young, sensitive and oily skin.

Cocoa-Butter Hand and Body Lotions

Cocoa-Butter helps soothe sore skin. Irritated skin may benefit from Chamomile and Marigold. Myrrh Oil aids healthy nail growth. To use on sore and dry skin.

Skin type: Dry skin.

Peppermint Foot Lotion

A rich, cooling foot lotion made with cocoa-butter and lanolin to soften hard skin, with Arnica, a well known herb for bruises, to soothe tired feet, peppermint oil to overcome foot odour and menthol to invigorate the feet and assist in counteracting chilblains.

Cocoa-Butter Suntan Lotion

Contains sesame oil which filters the sun's ultra-violet rays, and black walnut leaves which slowly colour the skin brown and aloe vera which produces a soothing effect on the skin, with a cocoa-butter base, absolutely nongreasy. To be used both before and after sunbathing, superb for maintaining the tan.

Freckle Lotion

Potassium chloride	1.2%
Borax	1.0
Potassium carbonate	3.7
Sugar	3.7
Glycerin	9.0
Rose water	20.0
Alcohol	10.0
Distilled water	51.0
Perfume	0.4

Make separate solutions of potassium carbonate and potassium chlorate and borax with a portion of the water. Dissolve the sugar in the remainder of water. Add the glycerin and rose water. Mix and then add the other solutions individually mixing before each addition. Add alcohol and perfume.

Caution: Do not mix dry potassium chlorate with organic substances.

Freckle Lotion

Acetic acid	3.0%
Concentrated lemon juice	10.0
Glycerin	6.0
Water	70.0
Perfume	1.0
Alcohol	10.0

Dissolve the concentrated lemon juice and the acetic acid in the water. Mix the perfume with alcohol and glycerin and add the solution to the lemon juice solution. Mix and filter.

5

Oral Hygiene Products

DENTAL CARE PRODUCTS are meant for keeping the dental struc
ture, healthy, strong and protected against any infection (oral).
These are also meant for keeping the enamel on teeth intact. These
products can be classified as normal and medicated dental preparations.

An ideal dental care product will remove dental plaque and tartar
and at the same time does no damage to the enamel of the teeth. A
tooth powder can be handled either by hand or with the help of a
brush whereas it is advisable to use a toothpaste or a gel, in conjunction with a brush. A medicated dental product should prevent
or protect dental decay.

FEEL

After the usage of dental product, it should leave a tingling effect in
the mouth; keep the oral cavity free from bad odour.

A general formulation of a dental product should consist of the
following ingredients.

- An abrasive
- A sweetener
- A foaming agent

- A polishing agent
- A preservative
- A cleansing agent
- A colouring agent
- A flavouring agent/
- A bleaching agent

TOOTHPOWDERS—GENERAL FORMULAS

Formula 1

Hard soap powdered	5 parts
Calcium carbonate precipitated	93.4
Saccharin soluble (Sodium saccharin is nowadays replaced by sodium cyclamate)	0.2
Oil of peppermint	0.4
Oil of cinnamon	0.2
Methyl salicylate	0.8
	100.0

Formula 2

Microcrystalline aluminium hydroxide	50 parts
Sodium lauryl sulfoacetate	2.0
Saccharin	0.1
Gum tragacanth	0.1
Water	47.8
	100.0

Formula 3: Ammoniated Tooth Powder

Dibasic ammonium phosphate	5.0 parts
Carbamide	3.0
Bentonite	6.0
Saccharin soluble	0.2
Menthol	0.2
Calcium carbonate ppt	84.2
Oil of peppermint	0.2

Oil of cinnamon   0.2
Oil of wintergreen   0.6
Dyponal (SLS)   0.4
                                       100.0

Formula 4: Chlorophyll Toothpowder

Dicalcium phosphate dihydrate   91.8 parts
Sodium copper chlorphyllin   0.2
Sodium lauryl sulphate   3.0
Sodium tripolyphosphate   3.0
Oil of peppermint   2.0
                                       100.0

Formula 5: Penicillin Toothpowder

Calcium carbonate (ppt)   94.27 parts
Tricalcium phosphate   0.2
Sulfocalmate   2.03
Menthol   0.2
Methyl salicylate   0.8
Oil of peppermint   0.2
Saccharin soluble (Sodium saccharin nowadays
is replaced by Sodium cyclamate)   0.3
Potassium penicillin, 500 units/   2.0
                                       100.0

Formula 6

Calcium carbonate   72.0 parts
Sodium bicarbonate   2.0
Tricalcium phosphate   15.0
Neutral soap   6.5
Saccharin   0.3
Essential oils   2.2
Penicillin 1,00,000 units   2.0
                                       100.0

Formula 7: Anti enzyme Toothpowder

Sodium N-lauryl sarcosinate	3.0 parts
Dicalcium phosphate dihydrate	94.2
Calcium phosphate dihydrate	0.3
Flavour	2.5
	100.0

Formula 8: Fluoride Toothpowder

Microcrystalline aluminium hydroxide	91.25 parts
Aluminium hydroxide (325 mesh)	5.2
Flavour	2.0
Saccharin soluble	0.25
Sodium fluoride	0.1
Sodium lauryl sulfoacetate	1.2
	100.0

Formula 9: Tartar Removing Toothpowder

Siliceous earth	95 parts
Erythrosine	0.1
Gluside USP	1.5
Cassia oil	0.4
Sodium benzoate	2.0
Clove oil	0.8
Menthol	0.2
	100.0

These toothpowders should not contain carbonate soaps or any alkaline materials. The active ingredient is sodium benzoate as it acts as a solvent on the tartar incrustations.

Formula 10: Antacid Toothpowder

Zinc oxide (Purified)	10.0 parts
Precipitated chalk	46.0

Tricalcium phosphate	20.0
Soda bicarbonate	6.5
Powdered soap (neutral)	6.5
Confectioner's sugar	11.0
Flavour of choice q.s.	
	100.0

Charcoal Toothpowders

Formula 11

Fine levigated wood charcoal (carbolignin) is the choice raw material.

Powdered charcoal	40 parts
Powdered cuttlefish bone	10
Prepared chalk	20
Heavy magnesium carbonate	20
Lactose	10
Clove oil	q.s.
	100.0

Formula 12

Activated charcoal (abrasive) obtained from vegetable sources	80.0 parts
Absorbent(Magnesium carbonate)	16.0
Foaming agent	2.0
Sweetener	0.1
Oil of wintergreen	0.6
Oil of peppermint	1.2
Oil of thymol	0.1
	100.0

Cocoa Toothpowder

Cocoa powder inhibits plaque formation and further, the combination of the cocoa powder and flavouring agent results in a taste pleasing, anti-plaque oral composition for dental use.

MANUFACTURE OF TOOTHPOWDERS

The manufacture of tooth powders is a relatively easy and simple operation compared to other dental products. The primary objective is the homogeneous distribution of all the ingredients without contamination by foreign substance or possible reactive equipment.

In case of tooth powders it is advisable to mix the ingredients in relatively small quantities first, before admixture with the reminder of the components. The flavour can be sprayed into the bulk of the powders during the mixing process, which is done in a ribbon type mixer.

Some tooth powders tend to form lumps are not free flowing. The process for overcoming this involves granulating the powders by drying slurries containing the very finely divided polishing agent, a detergent and a small amount of binder. The dried product is then mixed with the flavouring materials. The size of the granules of the tooth powders should be such that substantially all are retained on a 100 mesh screen but pass through a 40 mesh screen.

Packing Dentifrices

Tooth powders are generally packed in metal cans with dispensing top and closed with a metal or plastic cap. The cans are generally made of tin-plated or chemically treated steel and may be coated internally with a suitable lacquer.

Quality Control

It is important to standardize all factors concerned with the manufacture of dentifrices to ensure uniform distribution of all ingredients. To do this, specifications and control procedures for the raw and packaging materials finished product, manufacturing procedure should be written and adhered to.

LIQUID DENTIFRICES/TOOTHPASTES

In addition to the ingredients presented in the preceding sections, dentifrices have been reported to contain astringents (zinc chloride), preservatives (dischlorophene benzoate) esters of p-hydroxy benzoate, formaldehyde, oxidizing agents (potassium chlorate, sodium perborate, urea peroxide, magnesium peroxide) magnesium hydroxide, glycono delta lactone and chloroform.

In many instances, the added ingredients were intended to enhance the efficacy of the dentifrice. Although evidence to support this efficacy is not available in literature.

The need for care in selecting ingredients intended to fulfil a specific purpose (such as preservatives) is again evident from reports where patients developed side effects due to the presence of some such sensitive ingredients.

Formula 1

Soap powdered	7.2 % or parts
Saccharin	0.20
Amaranth (Solution)	1.0
Oil of cinnamon	0.6
Oil of clove	1.2
Oil of peppermint	0.6
Alcohol	29.2
Distilled water	60.0
	100.04

Formula 2

Sodium alginate	2 % or parts
Sodium fluoride	0.1
Water	66
Crystalline aluminium hydroxide	25
Flavouring agent, detergent, colour	q.s.
Dilute hydrochloric acid	4.2
Water to make	**100**

Formula 3

Chalk	39.5 % or parts
Water	32.5
Glycerol	20.0
Soap	6.3
Gum	0.4
Saccharin	0.1
Flavouring oils	1.2
	100.0

Formula 4

Calcium carbonate	45.7 % or parts
Starch	7.0
Glycerol	28.2
Water	14.4
Sodium benzoate	2.2
Flavouring material	1.3
Sodium salt of sulphuric acid ester of fatty acid monoglyceride	1.2
	100.0

It is a representative formula of a toothpaste in which calcium carbonate is used as the abrasive, starch as binder and the sodium salt of sulfated coconut monoglyceride as the surface-active agent.

Formula 5

Dicalcium phosphate	42.5 % or parts
Sodium coconut monoglyceride sulfonate	2.0
Water	28.35
Glycerol	25.0
Irish moss extract	1.45
Saccharin	0.1
Flavouring agent (mint)	0.6
	100.0

This is a toothpaste, wherein the abrasive is dicalcium phosphate, and the binder is Irish moss extract.

Formula 6

Calcium carbonate	35.5 % or parts
Tricalcium phosphate	4.3
Glycerin of starch	31.4
Magnesium hydroxide	3.8
Powdered white neutral soap	0.9
Potassium soap	0.73
Gum tragacanth	0.11
Propylene glycol	2.26
Flavour (peppermint oil/spearmint oil/ menthol)	0.8
Distilled water to make	**100.0**

Formula 7

Tricalcium phosphate	38.66 % or parts
Urea	13.0
Dibasic ammonium phosphate	3.0
Sorbitol	14.5
Glycerol	10.0
Water	16.64
2,2-dihydroxy 5,5-dichloro diphenyl methane	0.25
Sodium lauryl sulfoacetate	2.0
Aminoacetic acid	0.34
Carboxy methyl cellulose	0.28
Flavouring oil	1.1
Saccharin	0.23
	100.0

Formula 8: Chlorophyll Toothpaste

Calcium carbonate	55.0 % or parts
Magnesium carbonate	5.0
Sodium lauryl sulfate	5.0
Gum tragacanth	4.5
Chlorophyll (water soluble) or Copper chlorophyll	0.5
Distilled water	q.s.
	100.0

This formula is a representative of a toothpaste containing water soluble chlorophyll derivatives. This formula utilizes calcium carbonate and magnesium carbonate as the abrasives, SLS as the surface-active agent and gum tragacanth as the binder.

Formula 9: Anti Enzyme Toothpaste

Calcium carbonate	12.1 % or parts
Dicalcium phosphate dihydrate	36.2
Sodium N-lauryl sarcosinate	2.0
Glycerol	30.6
Water	15.3
Irish moss	1.0
Sweetening agent, flavour, preservative	2.8
	100.0

Formula 10

Microcrystalline aluminium hydroxide	38.2 % or parts
Aluminium hydroxide (325 mesh)	5.3
Sodium alginate	1.0
Sodium fluoride	0.1
Glycerol	15.3
Sorbo (70% aqueous sorbitol)	15.3

Water	20.8
Colour	q.s.
Saccharin	q.s.
Soluble flavour	1.2
Sodium lauryl sulphate	2.2
Orthophosphoric acid to pH 7.3	0.6
	100.0

Formula 11: Milk of Magnesia Toothpaste

Milk of magnesia	24.0 % or parts
Magnesium carbonate	10.0
Precipitated chalk	18.0
Soap powdered (neutral/white)	1.75
Glycerin of Starch	12.0
Glycerin	12.0
Water	20.0
Flavour	0.5
Methyl paraben	0.1
Mineral oil heavy	1.5
Saccharin	0.15
	100.0

MANUFACTURE OF LIQUID DENTIFRICES

The equipment required for the manufacture of liquid dentifrice is conventional, consisting of mixing and storage tanks, filtration units, glass lined or stainless steel tanks. In some cases manufacturing involves simple mixing of all the ingredients. In those instances wherein hydrophilic colloids are included to increase viscosity, the material is homogeneously dispersed in a portion of the solvent before mixing with the remaining ingredients. Flavours are generally dissolved in alcohol to which glycerol, aqueous solution of colouring matter, sweeteners and detergents are added.

Manufacture of Toothpastes

There are two general methods utilized in the manufacture of tooth-pastes.

1) The binder wetted with humectant is dispersed in the liquid portion containing the saccharin and preservative and allowed to swell, to form a homogeneous gel. Heating and agitation are used for acceleration of swelling. This gel is pumped into a suitable mixer such as Abbe Dispersall and the solid abrasive is slowly distributed with stirring until the whole mass is homogeneous, it can also be done the other way. With the abrasive weighed into the mixer and the liquid gel is added slowly with agitation until a homogeneous paste is formed. The flavour and detergent are added last and distributed uniformly. Excessive aeration particularly with the addition of the detergent should be avoided. The paste can then be milled, deaerated and tubed. In the second method, the binder is premixed with solid abrasives and introduced simultaneously with an aqueous solution of the humectant, preservative and saccharin into a suitable day dough type mixer. After mixing to a homogeneous paste, the flavour and detergent are added. However, since heat is not used in this method, it must be ensured that the binder can be sufficiently swollen during the mixing process so that the paste is uniform and of desired consistency.

2) Alternatively a toothpaste manufacturing procedure recommends moderate heat to accelerate swelling of binder. The glycerol, Irish moss, water, preservative and saccharin are mixed and heated to 100°F to form a gel like mass. The sarcosinate is then added with agitation to form a homogeneous mass after which calcium carbonate and dicalcium phosphate dihydrate are incorporated. After cooling and flavouring the mass is milled, deaerated and strained to produce a smooth homogeneous paste.

As in the case of toothpowders, adequate care must be taken to ensure that the manufacturing equipment does not interact with the paste thereby changing its colour or taste. Stainless steel and glass-lined equipment are best suited for this purpose.

Packing of Dentifrices

First tin collapsible tubes were made by Frenchman 'Richard Alive' during the 1950s, but due to scarcity of tin lead tubes with tin coating were used. In case of less than 10% tin coating on lead there is a danger of lead being absorbed by the paste. Hence for protection, wax spraying was done to prevent any such danger. Subsequently aluminium tubes were used.

In view of the possibility of tube corrosion or metallic contamination of the packaged product, it is important to establish the acceptability of a selected tube and protective coating. The closure of the tube is also an important aspect to consider. Formerly metal caps were used, but moulded plastic caps are currently in vogue. The inside of the cap has a plastic or a cork lining to seal the mouth of the tube. Recently plastic collapsible tubes have been introduced for specific advantages.

GEL TOOTHPASTE

A gel toothpaste is a gas-free or substantially gas-free viscous, extrudible paste or gel dentifrice comprising a polishing agent, a gelling agent and a vehicle and mixing with such dentifrice bubbles of gas of a size varying from 0.1–4 mm in diameter or containing 2–100 such bubbles per cubic centimeter of dentifrice. It is of a viscosity sufficient to maintain the bubbles therein. The clear gel dentifrices include polishing agents, gelling agents and vehicles, along with an anionic detergent or a foaming agent. Other adjuvants usually present are colour, flavour, antibacterial, preservative, buffering agents and an insoluble gas, i.e., a gas which will not dissolve objectionably in the dentifrice medium, although some of the gas may already be present in a dissolved state to create the dispersed bubble or sphere effect in the dentifrice.

Among the most useful polishing agents are complex aluminosilicates such as sodium aluminosilicate and silica xerogels, which are often partially hydrated (20%).

Gelling agents which may be useful to gelate or thicken the dentifrice include the natural and synthetic gums and gum-like materials such as sodium carboxymethyl cellulose, hydroxyethyl carboxymethyl cellulose, polyvinyl pyrrolidone, tragacanth, bean gum, starch glycolates and carbopol 934, 940, bentonite and other natural clays, proteinaceous materials either animal or vegetable derived, synthetic inorganic clays such as the silicated clays known as Laponite CP and SP to name a few.

The gelling materials employed are gelable with polyhydric alkanols such as glycerol and sorbitol and with water and lower alkanols.

Gel Toothpaste Formula

Glycerin	30.0%
Sorbitol (70% aqueous solution)	33.0
Laponite SP	2.0
Sodium aluminosilicate	20.0
Sodium N-lauroyl sarcoside	2.0
Flavour (essential oils)	1.0
Synthetic sweetener (saccharin)	0.1
Colouring solution (1% aqueous, green dye)	1.0
Sodium monoflourophosphate	0.8
Water	10.1

The Laponite SP, flavour, sweetener and colouring agent are mixed with approximately 1/3 of the glycerin and 1/3 of the sorbitol plus 1/2 of the water and a vacuum of 700 mm Hg is applied for 10 minutes. Then, 1/3 of the glycerin and 1/3 aluminosilicate and sodium monofluorophosphate and similar vacuum is applied to it for the same period of time to remove any entrained air. The sodium N-lauroyl sarcoside is next mixed in the remaining glycerin and sorbitol. The material is heated to 50°C and held for 5 hours without the application of vacuum or for 10 minutes with the same vacuum previously mentioned.

Then the polyhydric alkanol-gelling agent portion is mixed with the vehicle-polishing agent-fluoride portion, at a temperature of 40°C with the application of 700 mm Hg vacuum for 5 minutes, after which the surface active agent mixture is admixed, using the same vacuum and holding it for about 10 minutes. The mixing is done in a Unimix mixer, equipped with Teflon scraper blades, which clear the walls of the mixture to within 0.2 mm, leaving only a very thin film of dentifrice thereon. The product resulting is essentially gas-free, containing less than 0.1% by volume of entrained air. The pH thereof is about 8. (Product pH's within the process are 5 to 9.) The product resulting is a visually clear gel dentifrice of attractive appearance.

MOUTHWASHES

Mouthwashes are liquids more aqueous in nature and are to be applied in the mouth. They are solid or liquid concentrates which are diluted with water just before use. These concentrates have an advantage in economy of manufacture, shipping and storage but are not accepted as popular dosage forms. Mouthwashes generally contain:

- Flavour
- Antibacterial compounds
- Penetrants
- Astringents
- Therapeutic or Preventive compounds
- Deodorants

Broadly mouthwashes are classified into:

1. Cosmetic mouth washes consisting of water and alcohol, flavour (essential oil) and colour. May also contain sometimes surfactants (to solubilize essential oils) and to help in penetration and cleansing of the mouth and teeth.
2. Mouthwashes with a primary function of removing or destroying the bacteria in the oral cavity (solutions)
3. Astringent mouth washes, which in addition to their direct effect on the oral mucosa, also serves the purpose of flocculating and precipitating proteinaceous materials so that it can be removed by flushing.
4. Mouthwash concentrates which are desired for use after dilution.

5. Buffered mouthwashes, which depend for their action primarily on the pH of the solution. Alkaline preparations may be helpful in reducing stringy saliva or reducing nauseous deposits by dispersion of protein.
6. Deodorizing mouthwashes, which may depends on antibacterial action or on other mechanism for their effect.
7. Therapeutic mouthwashes, which are formulated for the purpose of relieving infections, preventing dental caries, or mitigating some other pathological condition of the mouth, teeth or throat.

The following are examples of a variety of mouthwashes.

Mouthwash Formula

Boric Acid	1.5 % or parts
Thymol	0.1
Eucalyptol	0.5
Methyl salicylate	0.1
Oil of thyme	0.03
Menthol	0.1
Alcohol	30.0
Water	67.67
	100.0

Dissolve boric acid in 60% of the water and other ingredients in 60% of the alcohol. Pour the aqueous solution into the alcoholic solution; add 2% of purified talc and allow the mixture to stand with occasional stirring for forty-eight hours. Filter the rest of the alcohol and water, and add caramel for colouring if a darker colour is desired.

Alkaline Mouthwash

Potassium bicarbonate	2.1 % or parts
Borax	2.0
Oil of sassafras	0.1
Thymol	0.05
Eucalyptol	0.1

Methyl salicylate	0.05
Tincture of cudbear	0.2
Alcohol	5.0
Glycerin	9.0
Water distilled	81.4
	100.00

Dissolve the potassium bicarbonate and borax in 50% of the water. Dissolve the oils in alcohol; add the alkaline solution, the tincture of cudbear, and the rest of the water. Mix thoroughly for two hours. Allow to stand for 48 hours. Chill and filter. Two percent of purified talc may be used to aid clarification.

Resorcinol Mouthwash

Resorcinol	5.0 parts
Zinc chloride	0.03
Menthol	0.5
Thymol	0.2
Eucalyptol	0.03
Camphor	0.03
Oil of peppermint	0.05
Alcohol	25.00
Glycerin	10.00
Water	59.16
	100.00

Dissolve resorcinol and zinc chloride in water, and the thymol, eucalyptol, wintergreen, menthol and camphor in the alcohol. Mix the two solutions together and add glycerin. Mix for one hour, chill and filter.

Salol-Thymol Mouthwash

Salol	0.3 parts
Thymol	0.15
Oil of lavender	0.03
Menthol	0.03

Benzoic acid	0.6
Glycerin	15.0
Alcohol	30.0
Water	53.89
	100.00

Dissolve salol, thymol, benzoic acid, and menthol in alcohol. Mix well. Mix tincture of cardamom with glycerin, add to alcohol, then add lavender oil and the water. Mix for four hours, chill and filter.

Astringent Mouthwash

Sodium chloride	2.5 parts
Zinc chloride	0.5
Menthol	0.05
Alcohol	0.6
Glycerin	10.0
Cinnamon oil	0.1
Water	86.25
	100.00

Dissolve the zinc chloride and the sodium chloride in water; add glycerin. Dissolve the rest of ingredients in alcohol. Mix the two solutions together and filter.

Aromatic Mouthwash

Eucalyptol	1.0 parts
Menthol	1.0
Clove oil	0.5
Wintergreen oil	0.01
Heliotropine	0.01
Chlorophyll, alcohol soluble	0.2
Water	50.28
Alcohol	47.0
	100.00

Dissolve all ingredients in the alcohol, and then add water. Mix for three hours. Chill in a cooling tank to 40°F, and filter.

Orange Mouthwash

Boric acid	2.5 parts
Chlorothymol	0.1
Methyl salicylate	0.6
Glycerine	10.0
Distilled water	86.8
Dark orange colour as desired	
	100.00

Acid Mouthwash No. 1

Benzoic acid	0.8 % or parts
Boric acid	1.6
Thymol	0.2
Oil of peppermint	0.3
Eucalyptol	0.2
Methyl salicylate	0.6
Chlorothymol	0.1
Alcohol	18.0
Distilled water	78.2
	100.00

Astringent Mouthwash No. 2

Zinc Chloride	0.1 parts
Menthol	0.05
Oil of cinnamon	0.13
Oil of clove	0.05
Solution of formaldehyde	0.05
Soluble saccharine	0.04
Alcohol	4.5
Distilled water	95.08
Colour, as desired	
	100.00

Cinnamon Type Mouthwash

Zinc chloride USP	16 parts
Saccharine USP	3
Formaldehyde USP	4
Hydrochloric acid USP (Conc. 36%)	2
Glycerin USP	13
Alcohol 38-B with menthol & cassia	32
Oil of cassia USP	
Amaranth powder	
Talc	15
Distilled water q.s. to make	**100**

Antiseptic Mouthwash

Boric acid USP	25.0 parts
Thymol	1.0
Eucalyptol	1.0
Menthol	1.0
Oil of gaultheria	1.0
Oil of thyme white	0.3
Benzoic acid USP	1.0
Floral extract of haptisia (Alcohol 48%)	8.0
Ethyl alcohol	25.0
Talc q.s.	
Distilled water q.s. to make	**100.0**

Dissolve the boric acid in water and the other ingredients in alcohol; pour the aqueous solution into the alcoholic solution, add the talc, and allow the mixture to stand 24 hours with occasional agitation. Then filter, retaining the first portions of the filtrate until it passes through clear. Add rest of alcohol to the filtrate and then sufficient water to make the product measure 100 cc.

Mouthwashes, Special Types

Pilocarpine—For Dry Mouth Relief

Pilocarpine, as a free base, occurs as colourless crystals with a melting point of 34°C and is soluble in water and ethyl alcohol.

Pilocarpine and its nitrate and hydrochloride salts, have long been known as parasympathomimetic agents. Ingestion of pilocarpine or its salts causes stimulation of the GI tract and stimulation of various glands, such as salivary glands, pancreas and mucosal cells in the respiratory tract.

Pilocarpine salts mentioned above in 10 mg dosages, three times a day temporarily restored salivation in patients suffering from dry mouth. But this method has a disadvantage as ingestion of pilocarpine may also produce undesirable side effects such as increased sweating, constriction of pupils, increased heart rate, increased gastric secretion in the stomach and increased bronchial secretions and to obtain relief the patient is required to follow a restricted diet.

It was found that pilocarpine as a free base or as a salt may be administered as a topical application in the form of a diluted solution to the mucosa lining of the mouth of patients suffering from dry mouth condition to produce long lasting relief from such conditions without undesirable systemic side effects. Pilocarpine hydrochloride as a dilute solution of 0.75 to 1% w/w in a mouth wash carrier is effective when gargled to produce long lasting relief from dry mouth condition. Since pilocarpine and its salts are bitter it is better to use a sweetened mouth wash carrier in order to mask the bitter taste.

Pilocarpine hydroxide	0.025 to 1%
Sweetener	1
Water	q.s.

This is specially useful to patients suffering from drug induced dry mouth, resulting from treatment with antidepressant, antipsychotic, antihypersensitive, and antiallergic medications.

For Removing and Preventing Dental Plaque

A method has been developed for removal of dental plaque and or dental calculus from teeth and the prevention of their formation. The process involves keeping the teeth in contract with a sufficient and effective amount of mouthwash. The mouthwash is a liquefied composition of an effective amount of fatty acid compound prepared from an unsubstituted, unsaturated fatty acid having at least one double bond, a liquid carrier, an effective amount of a buffering agent and an effective amount of ethanol. The pH of the liquefied composition is between 8 and 11.

A flavourant or flavourants, in small amounts can be added, which when mixed with saliva forms a protective film.

There is an oral formula containing proportions of ingredients resulting in the formation of a tenacious protective film when the rinse is mixed with saliva in the mouth the protective film produced by repeated (daily) use of such oral rinse formula has been found to give teeth a high luster or shine, reduce plaque attachment, diminish calculus attachment, gingival caries, reduce the discomfort of apthous ulcers and promote the healing of such ulcers as well as of cuts and abrasions in the mouth, and reduce and control sensitivity around the crowns and roots of teeth.

Sodium oleate	5
Ethanol	15
Disodium hydrogen phosphate to adjust to pH 8-11	q.s.
Distilled water to make	100

This mouthwash is particularly effective in preventing cavities or caries around metal teeth braces.

Zinc Sulphate Plus Ascorbic Acid

All the causative factors in the etiology of a healthy oral conditions are not known. It is known however that a reduction in the amount of zinc ions or in the amount of ascorbic acid available to nourish the oral tissue adversely affect their physiological tone. However, how much of this is the result of enzymatic, microbial and other factors has not been determined. It was clinically observed that sometimes the oral tissue becomes inflamed and susceptible to bacterial attack.

It is known that deficiency in diet of ascorbic acid or zinc renders the gingival cavity more prone to bacterial attack. However, excess amount of ascorbic acid or zinc in diet does not necessarily increase their content in the saliva or have a beneficial effect on the oral tissues. Nevertheless, it has been found that a combination of zinc ions and ascorbic acid provides a therapeutic composition, which improves the physiological tone of oral tissues apart from providing an effective solution against oral mocroflora responsible for plaque

Ethyl alcohol (95%)	20 parts
Zinc sulphate	2
Ascorbic acid	2
Glycerin	10
Water q.s.	100

This type is useful in gingivitis and periodontitis.

Formula to Form Protective Dental Film

Sodium fluoride	22 parts
Menthol	0.1
Oil of cinnamon	0.13
Oil of clove	0.05
Sodium saccharin	0.08
Ethyl alcohol	6
Distilled water q.s.	100

DENTAL CARE SPECIAL PRODUCTS

Denture Cleaners

Dispersion of dicalcium phosphate in montmorillonite clay has long been known as a suitable polishing agent especially for relatively soft surfaces. It finds application as a polishing agent for dentifrices, silver furniture and allied items.

One problem in using dicalcium phosphate is that it is difficult to create the sufficiently fine particle size necessary for its potential uses.

A method has been provided by researches to reduce the particle size of dicalcium phosphate by a process of utilizing sodium montmorillonite clay to form a stable suspension without effecting the basic physicochemical properties of the negatively charged montmorillonite platelet.

Another object is to provide a suspension of finely divided dicalcium phosphate in the finest particle size range free of any tendency for crystal growth.

When dicalcium phosphate is processed as described herein with a sodium montmorillonite clay, it has been found to create an extremely effective and useful product for polishing and cleaning.

Denture Cleansing Powder

Sodium tripolyphosphate	20.0 parts
Aerosil (finely divided SiO_2)	0.5
Sodium lauryl sulphate	0.3
Peppermint oil	0.5
Sodium saccharin	0.1
Urea	78.0
Sodium benzoate	0.6
	100.0

Effervescent Soak

In this category of denture soak formulas, some contain oxidizing agents as the source of effervescence, others a carbonate as the source from which carbon dioxide is liberated, and still others utilizing both oxidizing agent and carbonation agents (a carbonate plus an acid capable of releasing carbon dioxide) in a single composition. One of the problems encountered with these formulations is the lack of stability, largely due to the hygroscopic nature of the ingredients.

It has been found that a stable and effective denture soak product can be formulated by the addition of solid acid anhydride to a composition comprising oxidising agents, an inorganic carbonate and an organic acid. Specific examples of acid anhydride include, boric anhydride, succinic anhydride, adipic anhydride etc., to name a few, containing at least 30% by weight of urea.

There exists a need for a cleaning agent which effects quick and complete cleaning of the denture from the food particles collected particularly on the plaque. It has been found that such a composition must contain atleast 30% by weight of urea based on the total composition apart from other requisite substances for a complete end product.

A composition with a particularly quick and intensive cleaning capacity is obtained when the composition contains about 48% by weight or more of urea, the upper limit being 80% by weight of urea.

Dentifrice Speckles

Speckled macroscopic visible particles for oral paste or powder dentifrice compositions may contain in addition to organic binder one or more functional and/or aesthetic components of the dentifrice. The incorporation of speckles into such dentifrices in addition to enhancing appearance, has an added advantage when a functional or active ingredient of a dentifrice is present in the speckling material. These functional components when distributed homogeneously through the dentifrice can be rendered less effective due to their tendency to react with other functional dentifrice components, particularly on ageing.

However, their inclusion as speckles tends to keep them isolated thereby permitting greater scope in formulations of dentifrice. The incorporation of functional components into speckles provides satisfactory stability of the component until release.

Preparation of Speckles

Glyceryl tristearate	99.5 parts
Oil soluble (chlorophyll) colour index 75810	0.5
	100.00

The glyceryl tristearate is melted and agitated with chlorophyll. The resultant molten green mass is heated to about 80°C, in a vertical shaft mixer having a large diameter-mixing blade for 2 minutes at 300 rpm until the mixture becomes homogeneous.

The aqueous dispersion of green spherical particles is filtered to recover about 90% of yield of green speckles and reserved for uniform distribution in an oral dentifrice.

Speckling material

Glyceryl tristearate	79 parts
Zirconium silicate	20
Chlorophyll	1

Formulations Containing Speckles

Toothpowder	% or Parts	Toothpaste	% or Parts
Calcium carbonate	46.997	Glycerol 99.3%	19.8
Magnesium phosphate	0.2	CMC	8.5
Trisodium pyrophosphate	0.3	Sodium benzoate	0.5
Dicalcium phosphate dihydrate	46.3	Tetra sodium pyrophosphate	0.3
Sodium N-buroyl sarcosinate	5.7	Magnesum phosphate	0.2
Mint flavour	q.s.	Water	19.9
Glyceryl tristerate	0.5	Calcium carbonate	5.0
Chlorophyll	0.003	Dicalcium phosphate	48.797
		Sodium N-lauryl sarcosinate	5.7
		Mint flavour	0.8
		Glyceryl tristearate	0.5
		Chlorophyll	0.003
	100		**100**

Two Layer Cleansing Tablet

A two-layered denture cleansing tablet adapted for the self acting cleaning of dentures in an aqueous solution are also available these days in market.

OTHER FORMULATIONS

Chewing Gums Containing Plaque Inhibitors

Dental plaque is a product of microbial growth originating from residual food in the mouth. Plaque accumulates on the teeth and in the oral cavity. It is removed to some extent by brushing of teeth. However, some areas in the mouth are unaccessible and are susceptible to plaque and eventually calculus growth.

Several compositions have been provided for the purpose of inhibiting or reducing plaque in the oral cavity. But the most preferred and effective vehicle is found to be chewing gum.

Chewing Gums for Improvement of Saliva Flow

A gum composition packaged in stick form, has been developed. It contains gum base, fillers, in water based emulsifiers with dissolved solids and salts. In only about 15 minutes of chewing a very substantial quantity of salt solution is brought into the saliva and oral cavity to bathe the teeth and gum.

Tablet Toothpowder

Some manufacturers have come forward with toothpowder in tablet form.

6

Hair Care Products

HAIR DRESSING ONE type or the other has always enjoyed fascination among men and women over the ages. The tombs of Egyptian Kings (3500 BC) show evidence of use of perfumed hair oils while the Romans and Greeks used available fats and oils for personal grooming.

Most of the products in the earlier days were home made or derived from secret recipes containing wines, herbs, animal and plant by-products.

By the end of the 19th Century, some popular brands were systematically developed in collaboration with technical men with the objective of providing cosmetic elegance besides consumer acceptance of the product.

The scalp normally secretes oil sebum which gives a protective coating to the hair and prevents loss of moisture, keeps the hair in place and provides luster. It also protects the hair and scalp form harsh atmospheric and changing conditions. Lack of sufficient flow of oil causes dehydration of hair which makes the hair brittle allowing it to split and break.

Animal, vegetable or mineral oils and water have been judiciously incorporated in hair grooming preparations in order to give an ideal product.

PRODUCTS OF A GOOD HAIR DRESSING

A consumer acceptable hair dressing product will give the hair good grooming, luster without greasiness, protection from the elements and hair conditioning to a certain extent.

Luster

For luster a glossy material is required, say an oil or a fat or a solubilized fat which is absorbed on the hair shaft.

Cohesion and Adhesion

For grooming cohesion and adhesion are two basic necessities to be fulfilled. In other words the product must adhere to the hair fiber and provide cohesion to avoid an unruly condition. In addition to these two properties the material incorporated must possess another important property—lubrication which allows easy combing.

Conditioning

Moisture is the most important feature of conditioning. The oil in water emulsion and to a lesser extent the water in oil emulsion can give the best moisturising condition. Lanolin, fatty acid amides and some long chained quaternary ammonium compounds act as emollients. These along with the fats and oils protect the hair form loss of moisture and thereby minimize damage by the elements.

Types of Hair Dressings

1. Brilliantines liquid and solid
2. Alcoholic lotions
3. Two layer system
4. Hair tonics
5. Gum based hair dressing sprays
6. Oil-in-water emulsions
7. Water-in-oil emulsions

Brilliantines

The main purpose of a brilliantine is to add a measure of grooming and to impart sheen to the hair.

When the natural oils of the hair are either deficient or removed it has a dull appearance. Oil is the answer for such a problem. Originally, vegetable and animal oils were used. Since these natural oils tend to get rancid, they were substituted by mineral oil. This oil is not absorbed by hair but provides a thin protective film due to this low viscosity and penetration. The heavier the oil, the better its grooming properties, but its spreadibility reduces. Deodorized kerosene can be used to dilute the heavy oil to improve spreadability and penetration and after evaporation of the volatile deodorized kerosene, the thin uniform layer of viscous oil deposited on the hair serves the twin purpose of good grooming apart from imparting high gloss.

Liquid Brilliantines

Let us look at some liquid brilliantine formulas.

Mineral oil light	100%	75%
Deodorized kerosene	—	25%
Color and perfume	q.s	q.s

After mixing thoroughly, filter the solution. The viscosity of the mineral oil will vary with the final effect desired, and the method of application. But a problem of perfuming arises here, although both mineral oil and deodorized kerosene are relatively odour free, because of low solubility of perfume ingredients. Resinous and crystalline materials are insoluble and may precipitate on standing. Since brilliance is of utmost importance the effect of ageing, sunlight and heat on the final product should be studied. "Coupling" agents can be used for clear brilliantine. Even a small percentage of vegetable oil, a fatty alcohol, a fatty ester or some non-ionic surfactant will usually solubilize perfume oil.

Here are some liquid brilliantines with natural oils which have better absorption than mineral oil.

Castor oil	80%	6%	15%
Almond oil	20	49	—
Olive oil	—	45	—
Deodorized kerosene	—	—	85
Colour and perfume	q.s	q.s	q.s

Procedure

Thoroughly mix the ingredients, dissolve colour and perfume and then filter bright. The natural oils are subject to rancidity and therefore require anti-oxidants like tocopherols, octyl, dodecyl, cetyl, and stearyl gallates.

The phenol derivatives like thymol, pyrogallol, p-chloro-m-cresol are also active preservatives.

Here are some more formulas containing a mixture of mineral and natural oils.

Mineral oil	99-80%	66%	80%	80%
Olive oil	1-20	—	—	—
Peanut oil	—	34	—	—
Almond oil	—	—	10	20
Castor oil	—	—	10	—
Colour and perfume	q.s	q.s	q.s	q.s

Procedure

Mix all ingredients and then filter bright. Lately lanolin has been figuring in brilliantine compositions. However, because of its low solubility in mineral oil it has been found that by using certain fatty acid esters such as isopropyl mysristate or palmitate having a coupling effect, its solubility can be increased. The addition of these esters yields an ideal product because of their miscibility, resistance to rancidity and absorption by hair fiber. Furthermore they impart emolliency as well as gloss.

Here are some formulas illustrating the use of fatty esters.

Mineral oil	75%	38.5%	—	—
Ethyl myristate	25	—	—	—
Castor oil	—	38.5	60%	—

Isopropyl myristate	—	23.0	—	—
Ethyl oleate	—	—	40	—
Methyl oleate	—	—	—	25%
Olive oil	—	—	—	75%
Colour and perfume	q.s	q.s	q.s	q.s

A large number of synthetic nonionic materials are available these days and the one major advantage they have is that of not being subject to rancidity. Since strong sunlight does have an adverse effect on the hair, the use of protective additions in brilliantines does make sense although mineral oil by itself may be classified as a protective since it reflects a portion of the ultraviolet light of the sun spectrum.

Solid Brilliantines

Pomade or solid brilliantine are more or less synonymous. Originally, a pomade was the residue of fatty material left from enfleurage process of extracting floral odours. On the other hand brilliantines are vegetable or mineral oils hardended to desired consistency by the addition or certain waxes.

Solid brilliantines are very useful for curly or kinky hair because they hold the hair neatly in place. Solid brilliantines are also popularly used on closely cropped hair. The lubricity is just about enough to permit uniform spreading and therefore proper grooming.

The products are opaque and the opacity is directly proportional to the wax content.

Here is a simple formula of solid brilliantine.

Formula 1

Stearic acid	23%
Mineral oil	77
Colour and perfume	q.s.

Procedure

Melt the two ingredients together at 70°C. Add colour and perfume. Fill the molten product into jars and cool slowly in a warm room for 12-18 hours. A marbled product which crumbles at the touch is obtained.

Paraffin wax looked to be an obvious choice to give body to mineral oil. However, it presented certain problems like crystallization, permitting exudation of oil and shrinking on cooling leading to the brilliantine curving away from the sides of the jar. With the incorporation of petrolatum it was found that these problems could be solved. Here are some formulas containing petrolatum.

Formula 2

Paraffin wax	20%	15%
Mineral oil	50	25
Petrolatum	30	60
Colour and perfume	q.s.	q.s.

Use of ozokerite, spermaceti or ceresin wax yield better results in controlling the sweating of mineral oil. Here are some formulas containing these ingredients.

Formulas

	3	4	5	6	7	8
Mineral oil	67%	42%	86%	56%	70%	75%
Ceresin	11	8	—	—	—	—
Lanolin	22	—	—	—	—	—
Petrolatum	—	50	—	—	—	8
Spermaceti	—	—	10	22	5	—
Bees wax	—	—	4	22	—	—
Paraffin wax	—	—	—	—	15	—
Stearic acid	—	—	—	—	10	—
Ozokerite	—	—	—	—	—	17
Colour and perfume	q.s.	q.s.	q.s.	q.s.	q.s.	q.s.

Procedure

Melt together all the materials at the lowest possible temperature.

Add perfume as cooling occurs and fill into warmed jars. Allow to cool slowly. Care must be taken now to work the materials, particularly when partially solidified, in order to avoid air bubbles.

Vegetable oils have been used in solid as well as liquid brilliantines. Formulas hereunder are suggested to indicate general proportions of oils to waxes.

Formulas

	9	10	11	12	13
Castor oil	50%	—	—	80%	44%
Almond oil	30	85%	—	—	—
Spermaceti	20	10	—	—	—
Cocoa butter	—	5	—	—	12
Coconut Oil	—	—	75%	—	—
Ceresin	—	—	25	—	—
Bees wax	—	—	—	20	—
Petrolatum	—	—	—	—	44
Colour and perfume	q.s.	q.s.	q.s.	q.s.	q.s.

The synthetic waxes and some of the high molecular weight polyoxyethylene derivatives could be used but care must be taken in formulation since they are very lipophilic.

The perfuming of solid brilliantines is easier than that of clear liquid preparations. The amount of perfume used in solid brilliantines in usually higher than in equal quantity of liquid brilliantine since it is less volatile in the more viscous medium and since a much less quantity of grooming material is applied to hair.

Alcoholic Lotions

Alcoholic lotions have been popular since the beginning of the century. By diluting viscous oils with alcohol good wetting action is obtained and with the evaporation of alcohol, a thin uniform layer of oil deposited on the hair. Moreover, the temporary tingling sensation of the alcohol is liked by many.

Castor oil is a ready choice among fixed oils because of its free solubility in all proportions in alcohol. Another material often used is glycerol.

Regular use of alcohol on the scalp is undesirable because it may act as a dehydrating agent. When relative humidity is low, moisture will be drawn from the scalp and hair, leaving it dry and brittle. The addition of water reduces degree of dehydration proportionately. Further, alcohol tends to dissolve or extract oil from the skin and the addition of oils somewhat retards this process. As mentioned earlier, glycerol is often used, either by itself or diluted with alcohol. This too cannot be recommended since it is extremely hygroscopic leading to further worsening of dehydration of hair or scalp which may even lead to dandruff formation. Several examples of formulas containing the ingredients discussed are given below.

Formulas

	14	15	16	17
Ethyl alcohol	50%	80%	60%	83%
Castor oil	50	20	10	15
Glycerol	—	—	30	—
Tincture of benzoin	—	—	—	2
Colour and perfume	q.s.	q.s.	q.s.	q.s.

Tincture benzoin is said to reduce stickiness of castor oil.

All the materials are mixed well at room temperature and filtered bright. Perfuming is a problem since castor oil develops a characteristic odour on ageing although use of deodorized castor oil may help to reduce the problem. The use of anti-oxidants is not common in this type of product but is surely worth investigating.

Glycerol, fatty alcohols, fatty acids and the newer synthetic oils have also been used in alcoholic solutions. Here are some examples.

Formulas

	18	19	20
Isopropyl mirstate	5%	—	—
Glycerol	5	40%	—
Ethyl alcohol	90	40	96%

Water	—	20	—
Oleic acid	—	—	4
Colour and perfume	q.s.	q.s.	q.s.

Newer non-rancidifying materials have emerged for the formulation of hydro-alcoholic hairdressings. They are polyalkylene glycols, and are very easily soluble in alcohol or water. Less greasy products can be obtained than with some of the other oils. Large quantities of water can be added than with castor oil, which reduces dehydration.

Further solubilization of other oils in alcohols has been attempted for instance with the aid of deodorized kerosene. Formula hereunder illustrate such a preparation.

Formula

Ethyl alcohol	54%
Olive oil	19
Kerosene (deodorized)	27
Colour and perfume	q.s.

There is a good number of surface-active agents available (in the market) these days which could be employed in the above formula.

Hair Tonics

The purpose of hair tonic may be categorised into 3 functions. (i) To cure baldness, (ii) To relieve oily or dry scalp and (iii) Preventing or curing dandruff. However, none of the tonics have achieved success so far as growth of new crop of hair once it has been lost is concerned.

In some cases loss of hair can be checked for reasons like wrong treatment or carelessness leading to clogging of sebaceous glands and hair follicles, dehydration of scalp, poor circulation or infection. Proper care, stimulation and prophylaxis create favourable conditions for a healthy hair growth, but cannot stimulate new growth on baldhead.

A hair tonic usually is a combination of – a sebaceous gland stimulant, a rubifacient and antiseptic. Rubifacients commonly used are chloral hydrate (2-4%), formic acid spirits (10-12%) quinine and its salts (0.1-1.0%), tincture of cantharides (1-10%) and such tars as cade, preen and birch (0.1-10%).

These materials must be very carefully incorporated especially in newer formulations with wetting agents because they increase the flow of blood in the skin and capillaries due to which reddening occurs. If used in higher concentrations it may lead to irritation or even necrosis. The formulation must be able to cater to a wide range of individual sensitivity.

The antiseptics generally used are phenolic compounds and derivatives although phenol by itself is not used due to its toxicity and irritation. Some important phenolic derivatives include p-chloro-m cresol, p-chloro-m-xylenol, o-phenyl-phenol, o-chloro-o-phenyl phenol, p-amyl phenol, chlorothymol, resorcinol, and β-naphthol.

Except resorcinol which is used upto 5% all the other antiseptics used are less than 1% because in higher concentration they may cause irritation.

Quinine and its salts, tincture of jaborandi, leaves of pilocarpine, resorcinol monoacetate, cholesterol, salicylic acid, ethyl alcohol, methyl linoleate, sulphur and lecithin are few claimed "stimulators" of sebaceous glands.

Hair tonic meant for only scalps are alkaline and generally astringent. Tannin merits mention in this respect. The formulas hereunder are divided into the following categories.

1. To retard loss to hair.
2. To control dandruff and alleviate oily scalp.
3. For dry scalp.

Formulas of Bay Rum Hair Tonics

	1	2
Jamaica rum	12%	10.0%
Ethyl alcohol	45	50.0
Oil of bay	2	2.0

Glycerol	5	—
Water	36	37.5
Oil of pimento	—	0.5

Procedure

Dissolve the oils in the alcohol, add the remainder of the ingredients, and mix well. Filter bright and bottle.

	3	4
Oil of bay	0.15%	0.25%
Oil of clove	0.15	—
Ethyl alcohol	60.00	65.00
Tincture of quillaia	10.00	—
Water	29.55	34.59
Ethyl acetate	0.15	—
Oil of cinnamon leaf	—	0.05
Quassia extract, solid	—	0.10
Hepataldehyde	—	0.01

Procedure

Dissolve the essential oils in the alcohol, and the ethyl acetate or heptaldehyde, dissolve the quassia extract in the water with heat, and mix with the rest of the materials, stir well and filter bright.

Formulations using resorcinol and its monoacetate normally contain a rubifacient, as demonstrated in formulas:

Formulas of Resorcinol Hair Tonics

	6	7	8	9	10
Resorcinol	5%	0.8%	0.3%	—	—
Resorcinol monoacetate	—	—	—	3%	2.5%
Tincture of capsicum	5	—	—	—	—
Chloral hydrate	—	1.5	—	—	—
Spirits of formic acid	—	—	—	20	—
Pine tar oil	—	—	2.7	—	—
Ethyl alcohol	85	80.0	—	70	93.0

Castor oil	5	—	—	7	—
β-naphthol	—	0.8	—	—	—
Sulfonated castor oil	—	16.9	—	—	—
Soft soap	—	—	0.5	—	—
Potassium sulfate	—	—	3.0	—	—
Water	—	—	93.5	—	—
Methyl linoleate	—	—	—	—	2.5
Cinnamon	—	—	—	—	2.0
Perfume and colour	q.s.	q.s.	q.s.	q.s.	q.s

Procedure

Add the tinctures and oils to the alcohol in which the resorcinol or resorcinol mono-acetate has been dissolved; then add perfume and colour. After stirring well, filter the batch to clarify.

Pilocarpine containing lotions can be prepared from either the alkaloid or the tincture of jaborandi, as indicated in formulas:

Formulas of Jaborandi Hair Tonics

	11	12
Tincture of jaborandi	5.0%	—
Tartaric acid	0.5	—
Ethyl alcohol	5.0	9.0%
Triple rose water	82.5	—
Glycerol	7.0	—
Pilocarpine nitrate	—	0.05
Tincture of cantharidine	—	0.95
Water	—	85.0
Glyceryl borate	—	5.0
Perfume	q.s.	q.s.

Procedure

Prepare same as previous formulations. Great care being taken to obtain a brilliantly clear product on filtration.

Formulas given below are said to stimulate the sebaceous glands and to relieve a dry scalp.

Formulas for Dry Scalp Tonics

	13	14	15
Ammonia water	1.5%	—	—
Sulfonated castor oil	9.5	—	—
Tincture of capsicum	0.8	—	—
Ethyl alcohol	88.2	87%	2.500%
Chloral hydrate	—	3	—
Castor oil	—	10	—
Potassium sulfate	—	—	5.500
Water	—	—	91.690
Hydrochloric acid	—	—	0.004
Glacial acetic acid	—	—	0.006
Pine tar oil	—	—	0.300
Perfume	q.s.	q.s.	q.s.

Procedure

Mix the ingredients well and filter to obtain a brilliant product.

In the case of an oily scalp and mild dandruff frequent shampooing and application of hair tonic especially formulated to control secretion of the sebaceous glands is recommended. Here are some formulas with astringents, stimulants, and cleansing agents.

Formulas for Oily Scalp Tonics

	16	17	18	19
Tannin	5.00%	—	—	—
Formladehyde	0.75	—	—	—
Water	83.75	32.5%	—	56.5%
Ethyl alcohol	10.5	—	70%	40.0
Bay rum	—	30.0	—	—
Rose water	—	24.5	6	—
Ammonia 26°C	—	5.0	—	—

Glyceryl borate	—	7.0	—	—
Tincture of capscium	—	1.0	—	3.0
Eau de cologne essence	—	—	10	—
Glycerol	—	—	4	—
Tincture of cinchona	—	—	4	—
Tincture of quillaia	—	—	6	—
Chlorothymol	—	—	—	0.1
Quinine sulfate	—	—	—	0.1
Benzonic acid	—	—	—	0.3
Perfume	q.s.	q.s.	q.s.	q.s.

Procedure

Dissolve the water-soluble materials in the water, and the oils, tinctures, and perfume in the alcohol. Then mix, stir well, and filter bright.

In order to stimulate blood circulation in the scalp and control the activity of sebaceous glands regular massage is invaluable. The inclusion of more antiseptics like the quaternary ammonium compounds or hexachlorophene along with wetting agents, which help in the penetration of dandruff scales facilitate the removal by subsequent shampooing. However, one has to be aware of the fact that surfactants inactivate the antiseptics.

Two Layer Lotions

Brilliantines which are greasy in nature have to be diluted with either water or alcohol. This results in a two-layer system. As a result of which the contents have to be mixed well before use by shaking the container. The breaking into two layers can be somewhat corrected by addition of emulsifiers to form an emulsion but only temporarily. On standing the preparation again breaks into two separate layers.

It must be noted that the specific gravities of each phase must be sufficiently different so that it is a clear cut separation over of wide range of temperatures without inversion occurring. Each phase should be coloured separately so as to enable complete solution of the dye without diffusion of colour when the phases are mixed.

Furthermore the perfume used should not cause turbidity or precipitation at the inter surface.

A simple and inexpensive formulation is a mixture of water and mineral oil with colour and perfume. Water can be substituted in part or fully by alcohol. This yields a product that dries quickly and prevents breakage of bottle on freezing.

Formula

	20	21	22
Mineral oil	50%	65%	50%
Water	50%	—	32
Ethyl alcohol	—	35	18
Perfume and colour	q.s.	q.s.	q.s.

The mineral oil in turn can be partly replaced by refined vegetable oil according to degree of grooming required. Antioxidants must be added to prevent rancidity. The two phases should be mixed thoroughly and allowed to stand with intermittent stirring and finally filtered to give a crystal clear product.

Here are some more formulas of two layer lotions.

	23	24	25	26	27	28	29	30
Mineral oil	—	50%	32%	—	80%	28%	—	—
Castor oil	—	16	—	—	2	—	38%	—
Olive oil	8%	—	—	10%	—	28	—	—
Deodorized kerosene	—	—	—	—	—	—	—	5%
Seasame oil	—	—	10	32	—	—	—	—
Almond oil	—	—	—	—	—	—	5	5
Ethyl alcohol	45	—	58	58	18	44	57	40
Water	47	34	—	—	—	—	—	50
Perfume and colour	q.s.	q.s.	q.s.	q.s.	q.s.	q.s.	q.s.	q.s.

There is always the possibility of incorporating tonics in any one of the phases. However, little work has been done for the type of product due to its poor demand. The use of fatty alcohols and fatty esters to obtain more emolliency and sheen could be explored and a permanent emulsion could be attempted with the help of newer emulsifiers that are available these days.

Gum Based Hair Dressings

Gum based hair dressing preparation are useful in holding unruly hair in place for a long time. They do not impart sheen or luster to the hair.

A variety of gums natural as well as synthetic are available. Tragacanth and Karaya are popular although seed mucilage has also been in use. These apart, pectins, Irish moss, gum arabic, gelatin, flax seeds, mucilage and water soluble shellac have also been used. But one big disadvantage with natural materials is the variation in quality and properties from time to time and source to source.

The gum content in hair dressings ranges from 0.5%-2% depending on the type of gum and viscosity desired. On standing the products usually tend to thin due to either microbial decomposition or enzymatic reaction. A good preservative is therefore a must.

Tragacanth gum is the most popular among gums used. But it does not disperse easily in water. To avoid air bubbles when the gum is dissolved in water, and which are very difficult to remove, it is best to wet the powder with some inert water soluble solvent like alcohol or glycerol. Moreover, this facilitates bubble free dispersion in water.

Karaya gum in equal concentration is more mucilaginous than tragacanth but disperses more easily in water resulting in a whiter product. Here again it helps to wet the powder with say alcohol. Karaya sometimes has a characteristic odour of acetic acid which may be removed with the addition of borax or a mild alkali.

The products based on gums leave a uniform flexible film or drying. This holds the hair in place. The drying process can be quickened with the addition of alcohol upto 10%. Exceeding this limit will result in precipitation of the gums.

Polyvinyl pyrrolidine, (PVP) an alcohol, is best suited for this purpose especially when plasticized with glycols, and sorbitol and water soluble lanolin derivatives or a copolymer or methyl vinyl ether and maleic anhydride for quick dry, nontacky products as well as aerosol packing.

Methyl cellulose and sodium carboxymethyl cellulose form a film on drying but because of flaking and unsuccessful attempts to plasticize them lead to their rejection.

Carbopol 934 or synthetic resin and triethanolamine salt too have shown promise in such preparations.

Polyvinyl alcohol, polyethylene glycols (such as the carbo waxes), soluble starches, and abietic acid esters are some more potential ingredients for the preparation of mucilaginous hair dressings.

Many preparations produce sheenless and brittle, fibers which crumble or flake on combing. They can be supplemented with plasticizers like castor oil, mineral oil and some polyols and compatible newer nonionics to prevent this. Lanolin derivatives which are water soluble provide excellent plasticity and emollience. Furthermore, the presence of a humectant is necessary to sustain the moisture to keep the film flexible.

Gum based products warrant careful selection of perfume otherwise discolouration will occur on standing.

The formulas presented below are some gum based preparations.

Formulas for Gum-based Hair Dressings

	31	32	33
Gum karaya	—	—	2%
Gum tragacanth, powdered	1.2%	1.0%	—
Ethyl alcohol	15.0	6.0	5
Glycerol	2.0	1.0	—
Castor oil	—	2,0	—
Water	81.8	90.0	93
Preservative	q.s.	q.s.	q.s.
Perfume and colour	q.s.	q.s.	q.s.

Procedure

First wet the gum with alcohol and stir slowly to expel the air. Add the glycerol, preservative, castor oil, and perfume, and add all the water at once. Continue to stir until all the gum is dispersed uniformly, Filter or strain and allow to reach its maximum viscosity upon standing for a few hours. The addition of castor oil aids in plasticizing the gum film and in preventing dullness.

The addition of mineral oil yields a creamy product, and due to emulsification by the gum, apparently imparts more sheen but the strength of the film is reduced. One such formula is given below.

Formula

	34
Gum tragacanth, powdered	1.0%
Isopropyl alcohol	2.0
Glycerol	4.0
Mineral oil, heavy	1.8
Water	91.0
Formalin	0.2
Perfume and colour	q.s

Procedure

Mix the powdered tragacanth with the alcohol, add the glycerol, perfume and the mineral oil, and then all the water. Stir until the dispersion is complete and allow to stand for several hours to thicken up before straining. The appearance of the product can be improved by running it through a homogenizer.

Formulas for Gum-based Hair Dressings

	35	36
Sodium alginate	1.25%	1.5%
Glycerol	2.50	3.0
Calcium citrate	0.10	0.3

Distilled water	96.15	86.2
Tincture of benzoin	—	4.0
Balsam of Peru, 25%	—	5.0
Preservative	q.s.	q.s.
Colour and perfume	q.s.	q.s.

Procedure

Add the sodium alginate to half of the water, then add the glycerol and perfume, and stir well to dissolve. In the meantime, dissolve the calcium citrate in the remainder of the water, when the alginate, solution is smooth, pour in the calcium solution. Agitate well and allow to "body up" by standing a few hours. The glycerol can be replaced by alcohol, if desired. The viscosity can be increased by adding more calcium citrate or citric acid. Resins have been added to modify the film and to give a more opaque product. Add the tinctures directly to the alginate solutions.

Oil in Water Emulsions

Oil in water preparations were known in Europe for years but came to be known in America only during World War II. In a way they owe their introduction to America to the War. Because during this time alcohols and oils were rationed and the manufacturers had to turn to other materials. These products were an instantaneous hit with the consumers not only because of their novelty but also because of the smooth and attractive finish, good pouring and uniform film of oils on the hair. The greasy feeling is very much reduced and the residue is easily washable.

Furthermore, in a product like this replacement of some raw materials by a large portion of water in preparation relieves the manufacturer of a chunk of production costs.

Stability of emulsion is important for good shelf life. There was a feeling that these emulsions were probably too stable as a result, of which the inner phase was not reaching the hair. But later, it was found that it may be partially true in the case of water-in-oil emulsions but not so in oil-in-water emulsions. On application the water is absorbed by the hair breaking the emulsion. This leaves behind a protective film of the oils and fats on the hair shaft, which provides good grooming and luster.

In an emulsion the interfacial area between oil and water being large the changes of rancidity and deterioration of the natural oils and fats are high. Hence the selection of preservatives and anti-oxidants play a very important role. Since there are different culture mediums, and since the possibilities of contamination are varied it is difficult to select one particular preservative. It has to be found by extensive trials to suit the situation.

The preservative must be water-soluble so that it will be present in aqueous phase. Great care is necessary during manufacture storage and filing of emulsions in order to avoid contamination. Two equally effective, mutually compatible preservatives should be in use alternatively to prevent possible build-up of resistance.

A good emulsion must possess cosmetic elegances, and should provide gloss, should be non-greasy and easily applicable. Bees wax is a good gloss enhancer. But it should be well noted that excess solid content leaves a white deposit on the hair. Also the use of number of ingredients with a wide range of melting points results in unstable emulsions particular in freezing condition.

Oil-in-water emulsions provides uniform wetting thanks to the emulsifiers, which lower the surface tension of the aqueous phase. With the addition of emollients such as lanolin fatty esters, fatty acid amides and lecithin to the oils, conditioning can be improved.

The selection of right materials in the oily phase for desired grooming is necessary. Here petrolatum and waxes not only serve the purpose of substantiating the emulsions but also improve grooming effect. Vegetable oils may be preferred to mineral oils to get a better feel of the residual oils.

Gums and synthetic materials like magnesium aluminum silicate or polyvinyl pyrrolidone have found use as fixatives and emulsion stabilizers.

A good formulation for a hair dressing emulsion considers the following factors: (A) Selection of an appropriate emulsifying agent, (B) Proper balance between the oil and water phases (C) Correct viscosity of both phases.

Thus lotions with low oil content tend to foam on application, while white film deposits are caused due to occluded air bubbles and large content of stearic acid or other solids. Creaming which is different from breaking can be corrected by balancing the density and viscosity of the two phases.

The incorporation of an "antagonistic" emulsifier lends greater stability to the emulsion. Or in other words to stabilize an oil-in-water emulsion small quantities of water-in-oil emulsifier such as lanolin an adsorption base, or a fatty alcohol may be added.

A simple emulsion formula consists of oil, water and a sufficient quantity of emulsifier.

Although there is a large number of emulsifiers available in the market, here we shall look into four general groups starting with soaps.

Triethanolamine stearate heads the list of soaps used although other fatty acid salts have also been used.

Triple pressed stearic acid is commonly used to produce the soap by direct neutralization. Other acids can also be used. Many oil-in-water emulsions tend to get viscous and freeze in low temperatures and stop flowing. The addition of small quantity of oleic or another unsaturated fatty acid to stearic acid rectifies the situation.

Formula

	37
Mineral oil	44.0%
Stearic acid	6.0
Water	48.5
Triethanolamine	1.5
Perfume and colour	q.s.

Formula

	38
Mineral oil	25%
Triethanolamine stearate	7
Water	65

Bees wax	3
Colour and perfume	q.s

The first formula gives a heavy product and it can be made thinner by altering oil water ratio as shown in the next formula. (No. 38)

When the soap is prepared in situ as in formula 37 the emulsion is thick due to ready and continuous availability of emulsifier at the oil water interface and very mild agitation is required for the formation of the emulsion during mixing.

To conclude, the above two formulas illustrate the fact that the larger the concentration of the dispersed phase the higher the viscosity of the resultant emulsion. As a result thinning of emulsion occurs on dilution with water.

The formulation hereunder gives a viscous hair dressing with good gloss properties due to high wax content. The incorporation of glyceryl monostearate imparts added smoothness to texture.

Formulas for Emulsified Hair Dressings

	39	40
Mineral oil	43.0%	40.0%
Beeswax, white	3.0	1.5
Stearic acid	2.4	3.5
Glyceryl monostearate	0.2	—
Water	48.7	49.0
Triethanolamine	1.2	1.5
Carnauba wax	—	1.0
Stearamide	—	1.0
Perfume in 33% carbitol solution	1.5	2.5
Colour and preservative	q.s	q.s

The basic formulas discussed above can be enriched, with the addition of lanolin as an emollient for the hair, glycerol, propylene glycol, or other polyols as humectant, to protect the emulsion from drying out when exposed to air and gums for fixative properties and to achieve greater stability. Vegetable oils, fatty acids, and esters, alkanolamides, lanolin esters and lanolin oils for emollience and better grooming. Some of the following formulas contain emollients and humectants just discussed.

Formula

	41
Mineral oil	10.5%
Lanolin, anhydrous	6.4
Stearic acid	4.7
Quince seed mucilage	3.6
Propylene glycol	14.0
Triethanolamine	1.8
Water	59.0
Perfume and colour	q.s
Preservative	q.s

Procedure

Heat the propylene glycol, triethanolamine, and water to 70°C and pour into the molten fats. When the emulsion is formed add the quince seed mucilage, and then the perfume. The preservative should be dissolved in the water phase.

Formula

	42
Carbowax 1500	12.0%
Propylene glycol	3.0
Carbitol	5.0
Stearic acid	5.0
Lanolin	1.0
Triethanolamine	2.0
Potassium hydroxide, 85%	0.1
Sodium alginate, 2%	4.0
Water	69.7
Perfume and preservative	q.s

In the second formula soaps other than triethanolamine soap either alone or together with the latter can be used. Further, mineral oil is completely replaced by carbowax 1500, which results in a less greasy emulsion and has good grooming properties. The procedure is the same as the first one.

Formula

	43
Stearic acid	2%
Cetyl and stearyl alcohol, 10% sulfated	2
Soft white paraffin	8
Cocoa butter	4
Mineral oil	50
Water	33
Borax	1
Colour and perfume	q.s

Procedure

Heat the borax and water and add to the hot fats at 75 to 80°C, with moderate stirring. Continue slow stirring until the emulsion is cool, adding the perfume at 45°C into the oil phase and the sulfated alcohol in the water phase to cause emulsification during the mixing process.

The above formulas contain a combination of emulsifiers, which yields a better product than the one with a single emulsifier. The partially sulfated fatty alcohols on their own are very good emulsifiers, stearyl alcohol affords body and oleyl alcohol lends softness to the cream. Triethanolamine sulfate instead of a sodium salt results in a soft cream or a thinner lotion. Furthermore partially sulfated fatty alcohols with a slightly acidic (pH) can be used. They protect acid mantle of the scalp. Citric or tartaric acid may be used in emulsions made with sulfated fatty alcohols.

As mentioned earlier an emulsion contains oil and water with sufficient emulsifier to give the desired viscosity and a stable emulsion. Here is one such formula and the one after it for better grooming properties.

Formula

	44
Cera emulsificans	3%
Mineral oil	20%
Water	77
Perfume and preservative	q.s

Procedure

Add the heated water to the melt of mineral oil and sulfate alcohols at 75°C and stir the mixture until it is cool. This gives a fluid lotion which, due to its low oil content, would be inadequate for controlling unruly hair.

Formula

	45
Stearyl alcohol	5%
Mineral oil	33
Petrolatum	10
Sodium lauryl sulfate 15% solution	52
Colour and perfume	q.s

Procedure

Melt the stearyl alcohol, mineral oil, and petrolatum together to 70°C, and run this mixture into the warmed sodium lauryl sulfate solution under agitation. Add the perfume at 45°C. Cetyl alcohol or the softer myristic alcohol may be used to replace the stearyl alcohol.

A combination of vegetable oils and fats with partially sulfated fatty alcohols gives less greasy hair conditioners as shown in formulas below.

Formulas for Hair Conditioners

	46	47
Cera emulsificans	2%	15%
Peach kernel oil	18	—
Beeswax	1	—
Castor oil	3	—
Lanolin, anhydrous	—	3
Citric or tartaric acid	—	1
Water	76	81
Perfume and preservative	q.s	q.s

Procedure

Dissolve the water-soluble materials in the water, heat to 75°C, and pour into the hot oil phase. Stir until cool and add the perfume at 45°C to 50°C.

The polyhydric alcohol esters of fatty acids notably glyceryl monostearate have been widely and successfully used to give stable and attractive emulsions. The self-emulsifying (S.E.) grade of glyceryl monostearate is very popularly used since it does not require an anxiliary surfactant such as soap to give a stable emulsion.

An emulsions requires upto 3% of glyceryl monostearate whereas 10% of it will give a cream. The emulsifier reduces greasiness of oils, provides good emolliency and is very compatible with other oils, and fats. Another great advantage it has is its easy use. All the ingredients, water and oils can be mixed together and heated molten (stage) and stirred. However, it (GMS) is prone to mould attack, and the use of preservatives is prerogative. Propylene glycol stearate in place of the glyceryl ester yields a soft cream. The number of esters of this type is large and only useful ones are touched here. Diglycol stearate and laurate are most often used after glyceryl monostearate.

Some useful esters, produced by reacting the fatty acids with polyethylene glycol (PEG) are being widely used. Either a hydrophilic or lipophilic character can be obtained. Some lotions and creams are suggested in the following formulas.

Formulas for Hair Conditioners

	48	49
Mineral oil	30%	32%
Tegin	6	10
Water	64	50
Besswax	—	3
Castor oil	—	5
Perfume and preservatives	q.s	q.s

Procedure

Melt together all these ingredients and agitate until cool. Add the perfume at about 45°C. At first, the emulsion will have a very characteristic gelled appearance, but upon, cooling this will thin out to a smooth emulsion.

The purpose of lotions is usually for hair-grooming whereas the creams, with a higher solid content are useful for hair conditioning and treatments.

Formula 48 is a soft cream. The addition of bees wax for luster and vegetable oils in place of mineral oil lends variety to formula 49.

As already discussed humectants are added to retard drying of a cream, and the addition of fatty alcohols and lanolin provides a twin action; viz., emolliency and auxillary emulsification. By using fatty acid condensation of sorbitol and manitol and their polyoxyethylene derivatives known as spans and tweens oil-in-water emulsions can be made with less than 50% water content and with the availability of a large number of emulsifiers it all boils down to the nature of fatty materials and oils used.

Two formulas, the first one a free-flowing hair dressing and the other a cream are given below.

Formulas of Emulsified Hair Dressings

	50	51
Petrolatum	6.0%	15%
Mineral oil	37.5	10

Lanolin	3.0	20
Bees wax	12.0	12
Arlacel 83	3.0	—
Arlacel 20	1.0	—
Span 60	—	5
Tween 20	2.0	—
Tween 60	—	5
Borax	0.5	1
Water	35.0	32
Perfumes and preservatives	q.s	q.s

Procedure

Heat the oil phase plus the emulsifiers to 70°C, and the aqueous phase to 72°C and slowly add the latter, with agitation. Add the perfume at 45°C and stir continuously until cool.

Water in Oil Emulsions

A water-in-oil emulsion was first introduced in the year 1864. This emulsion simply consisted of Almond oil and lime water and had to be shaken well before use. Although there have been subsequent developments, these type of emulsions could never achieve the same type of stability as oil-in-water emulsions. Yet despite their tendency to break into two separate layers of oil and water. they have been widely accepted because of their creamy texture and ability to provide excellent grooming and sheen.

The film that is formed on the hair is water resistant and this property makes it popular with regular swimmers and is very useful in rainy climates.

Generally, humectants are not required for these type of formulations. The outer phase consists of the oily phase. This is what renders them light as a result of which the product can be used for treatment as well as hair grooming. Formulas containing one or more of the above mentioned ingredients.

Formulas of Hair Creams

	52	53	54	55	56
Tegin	12%	9.0%	13.5%	12%	3.0%
Mineral oil	2	25.0	8.5	2	8.0
Lanolin	4	—	3.5	10	—
Cetyl alcohol	—	1.5	—	—	—
Bees wax	—	1.0	1.5	—	—
Triethanolamine stearate	7	—	—	—	—
Water	75	59.0	59.5	68	78
Glycerol	—	4.5	4.5	3	5.0
Cholesterol esters	—	—	9.0	—	—
Spermaceti	—	—	—	5	2
Petrolatum	—	—	—	—	3.0
Stearic acid	—	—	—	—	1.0
Perfume and preservative	q.s	q.s	q.s	q.s	q.s

Diglycol stearate and laurate can freely replace glyceryl monostearate as shown in the following formula.

Formulas of Hair Creams

	57	58	59
Diglycol laurate	—	—	14%
Diglycol stearate	7%	8%	—
Mineral oil	20	—	36.0
Water	73	60	50
Castor oil	—	16	—
Almond oil	—	16	—
Perfume and preservative	q.s	q.s	q.s

Formula

	60
Stearyl alcohol	0.6%
Absorption base	2.5.

Diglycol distearate	6.0
Isopropylamine	0.3
Mineral oil	32.0
Stearic acid	0.6
Water	58.0
Perfume and preservative	q.s

In formula 60 the isopropylamine strengthens the diglycol stearate as an oil-in-water emulsifier while the absorption base plays the role of an antagonistic auxiliary emulsifier. This combination is often used to prevent an emulsion from breaking during application. An antagonistic emulsifier is there to remove the watery feel of the emulsion, while it is being rubbed into the scalp.

As a result the oil-in-water emulsion instead of separating can invert in a multiphase one and the smooth feel is relained till the end. Some formulas with polyethylene glycol (PEG) derivatives.

	61	62	63
PEG 300 monostearate	—	—	10.0%
PEG 400 monostearate	6.0%	4%	—
PEG 400 monolaurate	0.5	1	—
Lanolin	1.0	1	—
Polyethylene glycol	2.5	—	—
Water	90.0	92	54.5
Propylene glycol	—	2	—
Bees wax	—	—	8.0
Stearic acid	—	—	8.0
Mineral	—	—	18.0
Triethanolamine	—	—	1.5
Perfume and preservative	q.s	q.s	q.s

Procedure

For formulas 61 and 62: Heat all the ingredients to 70°C to 75°C stir well, and cool with moderate agitation.

Procedure for formula 63: Pour the triethanolamine water solution into the heated oils at 70 to 75°C. Add the perfume at about 45°C as usual.

The water-in-oil emulsions are more sought after by men folk rather than woman probably due to their high grooming property and oily feel.

Several types of emulsifiers as well as combinations of emulsifiers have been used to prepare water-in-oil emulsions. Beeswax and borax have been in use for a long time though polyvalent soaps are the oldest.

Calcium oleate and stearate prepared from lime water were originally used as emulsifiers before being replaced by a more stable magnesium salt. Use of a combination of water-in-oil emulsifiers has been found to be more effective rather than a single one.

The first formula is a simple water-in-oil emulsion. But is not very stable. The formulas following it contain beeswax or other waxes to add gloss; further the stability and emolliency of the emulsion are increased with the addition of vegetable oil instead of mineral oil and lanolin.

Formula

	64
Mineral oil	49%
Stearic acid	1
Lime water	50
Perfume and preservative	q.s.

Procedure

Melt the stearic acid in the mineral oil and add the lime water after heating to about the same temperature. Stir until cool and add the perfume at about 45°C.

Formulas of Emulsified Hair Grooms

	65	66	67	68
Mineral oil	45%	—	—	48.0%
Oleic acid	12	20%	10%	—
Beeswax	2	1	1.5	2.5

Lanolin	2	0.5	—	—
Lime water	19	33.5	53.8	—
Saccharated lime water	20	5	—	—
Olive oil	—	40	32.7	—
Magnesium sulfate, 25%	—	—	2.0	—
Stearic acid	—	—	—	1.0
Absorption base	—	—	—	5.0
Petrolatum	—	—	—	12
Magnesium oleate	—	—	—	2.5
Water	—	—	—	29.0
Perfume and preservative	q.s	q.s	q.s	q.s

Procedure

Add the aqueous phase at 70°C to the heated oils at 75°C, with moderate agitation, and stir until cool. In formula 67, add the magnesium sulfate solution after the lime emulsion has been prepared, and then stir the emulsion until cool.

The formula below is based on the beeswax borax emulsifying system.

Formula

	69
Beeswax, white	2.5
Mineral oil	62.55
Water	34.8
Borax	0.15
Perfume	q.s.

Procedure

Melt the wax in half of the mineral oil and add the hot water. Then add the remainder of the oil. Homogenization is of general value with this formula.

The following formula of a hair cream utilises a combination of emulsifiers.

Formula

	70
Mineral oil	40.0%
Petrolatum	19.4
Beeswax, white	17.6
Oleic acid	0.4
Lanolin absorption base	0.8
Borax	0.4
Magnesium sulfate	1.0
Sodium hydroxide	0.4
Water	20.0
Perfume and preservative	q.s

Procedure

Dissolve the magnesium sulfate and the sodium hydroxide in half of the water and the borax in the other half. Mix the two aqueous phases and add to the heated oils, both at 65°C. Stir until cool to 45°C then add the perfume and pass through a homogenizer.

Absorption bases formulated from lanolin, lanolin alcohols, cholesterol or cholesterol esters in combination with mineral oil, petrolatum and waxes hold a large quality of water to form water-in-oil emulsions. In fact, creams are produced rather than lotions, the absorption bases play an important role in imparting high grooming properties, stability, gloss and finish (fine texture) to the creams.

In short, a stable water-in-oil emulsion, simply consists of a right combination of absorption base and water.

The addition of bees wax and mineral oil produces a thinner emulsion with more gloss as illustrated in formulas given below.

Formulas for High-gloss Hair Emulsion

	71	72
Absorption base	63%	6.00%
Beeswax	—	3.00%
Mineral oil	—	60.25

Glycerol		—	1.50
Triethanolamine		1	0.25
Water		36	29.00
Perfume		q.s.	q.s.

Procedure

Melt the oil phase and heat to 70°C to 75°C and slowly pour the hot triethanolamine solution into the oil phase, with good agitation. A polyphase emulsion is formed first, but upon continued agitation and cooling, this changes substantially to a water-in-oil emulsion, and no further water can be added. Homogenization will increase the stability of the emulsion.

Arlacel 83 (Sorbitan sesquioleate) can be used in any of the system discussed and it gives stable lotions with low viscosity with 40%-50% water content. The formulas below show its use along with other emulsifiers.

Formulas of Low-viscosity Hair Emulsions

	73	74	75	77
Mineral oil	45.00%	37.5%	36.5%	33.0%
Petrolatum	8.00	7.5	8.0	—
Beeswax	3.00	2.0	2.0	3.5
Absorption base	7.00	—	—	—
Arlacel 83	4.00	3.0	3.0	—
Lanolin	—	3.0	0.5	4.0
Lanolin esters	—	—	—	10.0
Ceralan	—	—	—	5.0
Zinc stearate	—	1.0	—	—
Water	32.25	45.5	49.7	44.0
Borax	0.75	0.5	0.1	0.5
Magnesium sulfate	—	—	0.2	—
Perfume	q.s.	q.s.	q.s.	q.s.

Procedure

Add the aqueous phase slowly to the oil phase at about 75°C, with moderate agitiation. Continue to stir while cooling, and add the perfume below 45°C. Homogenization, although not necessary, will greatly add to the shelf stability of these water-in-oil emulsions. Care must be taken to carry this out at a given constant temperature range, lest the viscosity vary from batch to batch.

In the above formulas the beeswax-borax system is in use with the addition of sometimes an absorption base or a polyvalent soap.

Hair Straighteners

The general tendency of an individual with straight hair is to crave for curly hair and one with curly hair for straight hair. However, only in the 1940's, the operation of hair straightening became as easy effective and safe as hair curling.

Cross-sectional examinations of hair shaft indicated: that straight hair appeared circular and curly or kinky hair appeared like a very oblate spheroid. Although this theory has been accepted, it is by no means uncontested.

Whatever it may be, the simplest and oldest method of hair straightening was by way of plastering down the hair with gums or resinous-fatty vehicles as shown in the following formulas.

Formula 1

Quince seed	3%
Ethyl alcohol	25
Water	72
Perfume and colour	q.s

Procedure

Soak the quince seed overnight in 50 parts of water. Strain the resulting mucilage through a cloth and add the balance of the water, alcohol, colour and perfume.

Formula 2

Gum karaya	1.5%
Ethyl alcohol	5.0

Glycol bori-borate	1.5
Water	92.0
Preservative, colour and perfume	q.s.

Procedure

Make a mucilage of karaya, then add alcohol, and water. Add the glycol bori-borate, colour and perfume prior to passing through a colloid mill.

Formula 3

Petrolatum	90%
Paraffin	10
Perfume and colour	q.s

Procedure

Melt petrolatum and mix with paraffin and add perfume and colour, as desired.

In formula 3 paraffin wax may be replaced by bees wax or ozokerite. Normally paraffin is preferred because it is less expensive. Ceresin is also used if a hard product is needed. Certain resins can also be added in this type of a product. The degree of tackiness is directly related to wax content. Above 20% of wax content yields to hard a product for easy application.

The product of formula 3 enjoys popular usage among male population having too short a hair for recent techniques of hair straighteners to deal with.

The three formulas mentioned above have some disadvantages. They do not effect hair chemically. Their efficacy is exclusively based on their sticky qualities.

When curly hair is not changed chemically, it returns to its normal state. Moisture enhances this process. Bearing this in mind pomades with water repellents like aluminium stearate were made. But they proved to be ineffective, as they could not shield the hair from water vapour. Here is one such preparation.

Formula 4

Petrolatum	95%
Aluminium distearate	5
Perfume and colour	q.s

Procedure

Melt the petrolatum and add the aluminium distearate, perfume and colour.

Mme C.J. Walker conceived the idea of straightening the hair by the physical rather than the chemical methods. A hot metal comb and petrolatum jelly were used and the procedure was called "hair pressing".

In this method, the hair is washed and dried completely. Then the petrolatum product is applied to the hair. The hot comb is then passed through the hair which as a result of which the hair is stretched.

Here the petrolatum or "pressing oil" acts as a conductor between the hair and the comb and serves as a lubricant enabling easy passage of the comb through the hair without sticking and pulling. This method is used by both men and women. The press is used to straighten the hair, which is the one and only procedure used for males.

The second press is employed by women. It is equivalent to the general procedure of hair styling and curling which fascinates a majority of them. This second press uses croquignole irons. The second press is solely used by women, as males seldom prefer waved hair.

Laboratory trials indicate that a cooling period is required between the first and second press for a long lasting wave. Care must be taken while using the hot comb and croquignole iron in order to prevent the singeing of hair. This method is widely followed procedure of straightening women's hair both in beauty shops and at home.

There are limitations as far as the hot press method is concerned. It does not insure a permanent wave set. Water and perspiration have caused embarrassments to persons using this method although **water repellents were** used but with little success.

The hot press method is to overcome resistance offered by very curly hair to hair styling. But a person with straight hair can derive some satisfaction by simply moistening the hair with water and curling on rods and drying.

A simple formula (No. 5) gives a product that is used with hot combs and the following one (No. 6) a little more complex.

Formula 5

Petrolatum	100%
Perfume and colour	q.s

Procedure

Melt the petrolatum and add the perfume and colour.

Formula 6

Beeswax	7.00%
Ceresin	3.00
Petrolatum	30.00
Mineral oil	30.00
Water	29.25
Borax	0.75
Perfume	q.s.

Procedure

Melt the oils and waxes on a water bath and bring to 70°C. Dissolve borax in water and bring to 72°C. Add the latter solution to oil-wax mixture, with rapid agitation at first. After all the water is added, continue agitation slowly, cooling to 55°C, at which time the perfume is added. Continue to cool with agitation to 40°C and fill into jars.

Formulas 7 and 8 represent yet another more recent type of hair straightener with caustic alkali.

The consistency of these creams is related to the quantity of stearate or stearic acid and oleic acid used. The alkali employed, sodium hydroxide is proportional to the base used. Normally it does not exceed 5-9%. The increase in alkali provides quicker action on the hair but at the same time greater caution is to be exercised in application.

This type of product with alkali is preferred by men who prefer straight hair and require no styling.

Formula 7

Sodium hydroxide	5%
Glyceryl monostearate	15
Glycerol	5
Water	75
Perfume	q.s.

Procedure

Heat the glyceryl monostearate, glycerol and water to 95°C. Cool to 60°C, and then add the sodium hydroxide, which has been dissolved in a small amount of water. Cool with agitation to 40°C and perfume, and fill. It is important that the sodium hydroxide not be added at too high a temperature; otherwise the glyceryl monostearate will be saponified.

Formula 8

Stearic acid	15%
Oleic acid	5
Glycerol	5
Sodium hydroxide	10
Water	65
Perfume	q.s.

Procedure

Dissolve the sodium hydroxide and glycerol in the water and heat to 90°C. Heat the stearic acid and oleic acid to 95°C and add to the water solution with agitation. Cool with agitation to 40°C, add perfume and fill.

It may be noted that high alkali content in formula 8 warrants caution on the label for the benefit of the users.

An example of one such label is given below.

Place a little petrolatum along the hairline and on the ears before starting to use. Apply the product to the hair above forehead. Be careful not to allow the straightener to drop on the skin. Comb the straightener through the hair in an upward movement away from the scalp. Repeat the combing until the hair becomes as straight as you desire it. When straight, wash the hair in running water to remove the hair straightener. Rinse the hair until the soapy feeling is gone. Then wash with shampoo thoroughly. Rinse with lukewarm water. Do not retain this rinse water; use fresh water for every rinsing. Be sure not to use hot water.

If the hair is not thoroughly washed, the hair will turn red quickly and may break off at the scalp. When the hair is completely and thoroughly washed, apply hair pomade. It is suggested that you take the hair straightener to a barber to apply if you don't know how. The manufacturer does not assume any responsibility for the results if improperly used. The majority of the people can use the straightener without any bad results. However, some, through careless handling, will burn and discolour the hair. This is a result of the action of the individual and not the straightener.

Thioglycolates

With the acceptance of thioglycolates permanent hair straighteners, based on this group of chemicals, appeared in the market. They were much higher priced due to greater ingredient costs.

However, the introduction of this type of a product was not smooth due to:

1. Preparations being very different from the previous ones.
2. The necessity to provide highly detailed instructions to the user which made the manufacturers job that much more elaborate.

The introduction of thioglycolate products into the market did not come through without resistance from some quarters. Some medical writers were critical of the use of thioglycolates in anticipation of possible damages it would do to the hair. However, these products weathered all storms and finally they were accepted at least for waving if not for straightening.

A formula for a hair straightener has to consider in particular, the vehicle. This is secondary in a cold permanent wave although both products contain more or less same ingredients and encounter similar manufacturing and packaging problems.

In the case of waving process the hair is curled on curlers whereas in straighteners it is necessary to have a viscous and adhesive base that will keep the hair somewhat straight. A practical method has not been found yet and the day it is found a major problem in hair straightening will have been solved. Periodic combing and stretching, causes for many complaints when not properly done, can be done away with.

A majority of thioglycolate straighteners available commercially are of oil-in-water type of emulsions. Viscous liquids are also available. Formula 9 represents such a preparation.

Formula 9

Glyceryl monostearate	15.0%
Stearic acid	3.0
Ceresin	1.5
Paraffin	1.0
Sodium lauryl sulfate	1.0
Distilled water	51.9
Thioglycolic acid	6.6
Ammonium hydroxide (26°)	20.0
Perfume	q.s.

Procedure

Mix the glyceryl monostearate, stearic acid, paraffin, and sodium lauryl sulfate in a kettle, with 40 parts of the water, and heat, with constant agitation, to 95°C. Then cool to 50°C, still under agitation. While this is being done, add the thioglycolic acid to the remaining water and to this solution slowly add the ammonium hydroxide. As considerable heat is generated, it is necessary to provide some method of cooling to keep the temperature below 50°C. For this reason, it is preferable to use ammonium hydroxide by 16 parts. Slowly add the aqueous thioglycolate solution to the stearate mixture, both being at 50°C. It is important that the thioglycolate solution is not allowed to go above 50°C; otherwise there will be a decomposition of the ammonium thioglycolate. Care should be taken that the thioglycolate solution is mixed in thoroughly as it is added. Cool the resulting mixture rapidly to 40°C, to avoid saponification of the glyceryl monostearate, add perfume, and package

The control of free ammonia and of thioglycolate content is most important. The final product is assayed for the thioglycolate ion and the assay expressed as percentage of thigolycolic acid. Any adjustment due to loss of water during manufacture should be made to bring the thioglycolic acid within the range of 6.4 to 6.8%. Likewise, the free ammonia should be determined and adjusted to lie between 0.8 and 0.9%.

As variations, in place of ammonium radical, sodium salts or amines, such as monoethanolamine may be used.

Many manufactures recommend the usage of a neutralizer with thioglycolate as well as permanent wave products. The most popular neutralizers are perborate or bromate oxidizing mixtures, provided proper packaging precautious insuring stability are taken. Some variations incorporate surfactants for greater solubility and penetration and others vary the pH suitable to their needs.

A liquid bromate preparation is described in formula 10.

Formula 10

Sodium bromate	14%
Propylene glycol	2
Lanolin, anhydrous	2
Sorbitan monopalmitate	1
Sorbitan trioleate	2

Sodium cetyl alcohol sulfate	3
Water	76
Perfume	q.s.

Procedure

Dissolve the sodium bromate in the water and propylene glycol and heat to 60°C. Add the mixture of the remaining ingredients, which have been heated to 60°C. Cool to 35°C, with agitation, add perfume, and package. Because of the high concentration of salt (sodium bromate), this product may have a tendency to separate over a period of time, especially in hot weather. It is therefore advisable to use a 'shake well' label.

There have been debates within the cosmetic industry on the value of neutralizers being included in hair straighteners as well as waving preparations. Nevertheless, hair straighteners can derive certain benefits with the incorporation of a neutralizer.

In order to insure optimum activity a greater deal of control is necessary in the manufacture of thioglycolates than some other cosmetic products. Further, all raw materials must be thoroughly checked for metallic content including the perfume and the free alkali to be uniform. Iron, manganese and cobalt are to be avoided as they may cause catalytic decomposition of the mercaptan. The raw materials used for thioglycolates must fulfil the following conditions as per the directive of the Toilet Goods Association.

Table I: Toilet Goods Association Specification for Thioglycolates

Ash	0.05% maximum
Thioglycolate	45 to 55% as thioglycolic acid
Dithioglycolate	2% max. as thioglycolic acid
Iron	1 part per million max. as Fe
Copper	1 part per million max. as Cu
Lead	1 part per million max. as Pb
Arsenic	1 part per million max. as As_2O_3

1. It is for this very reason manufacture of thioglycolate products in glass, plastic, earthenware or stainless steel is a prerogative. Attention must be paid to values and other contact parts.
2. Temperature is another important factor to be considered. Heat, due to exothermic reaction, is generated during neutralization of thiglycolates and care must be taken of the product.
3. The product must not be exposed to air during manufacture or packaging to protect it against excessive oxidation.
4. Control of free alkali in the thioglycolate concentrate before mixing with the base and throughout the manufacture and packaging is a must.
5. A titration should be done to test whether the product has been properly packaged. Proper selection of packaging material to insure air tight and tamper proof sealing. Caps must be tested for reaction with the contents.

Collapsible tubes were used but not without trouble. If they are used perfect lining of tubes is indispensable. Polyethylene does not serve the purpose because of the high permeability of the plastic.

Neutralizers as with thioglycolate products, need utmost care in handling.

Now to turn to the powders, bromate and perborate must be packaged under dry atmosphereic conditions in air-tight and moisture proof containers. Most products of this type are packaged in aluminium foil coated with cellulose acetate. pH and oxidizing capacity need to be tested before and after packaging.

Method of Application

The use of thioglycolates should be proceeded by a careful study of instructions that go along with the methods of application. For example:

- Do not use hair straightener if you have abrasions or cuts on your scalp. Wait until they have healed before straightening your hair.

- Do not straighten your hair if you are sick or under the care of a doctor without consulting him first, since your hair is affected by your physical condition.
- Do not substitute hot water when directions call for warm water.
- Do not use the treatment unless you use all of the neutralizer in the kit. You must use the neutralizer as directions state.

For those with damaged hair, here are further instructions:

- Do not use hair straightener if your hair has been abused. If you have used hot combs, marcel irons, harsh lye straighteners, dyes, or bleaches, you may have abused your hair. For this condition, we recommend the use of hair conditioner for about two weeks or longer before straightening, to help recondition the hair.

The use of comb on the scalp which is softened due to the application of thioglycolate needs to be done with care lest irritation or abrasion leading to infection may occur. If it is a shampoo type hair straightener, it should be prevented from getting into the eyes during its application.

For hair straightening, the hair must be thoroughly cleansed of greasy oily, resinous films before application of the thioglycolate product if satisfactory results are desired. After the initial treatment combs or lye products are discouraged. Furthermore, a limited application or use of thioglycolate products is recommended be it for waving or straightening particularly for the latter purpose. This is limit being 3 or 4 times a year.

However hair straighteners in general are designed for the purpose, for which it is considered advisable that the reaction of the hair to the straightener rather than for texture of the hair, be the determining factor during the processing period.

Subsequently several new developments and formulations have surfaced in hair straightening. The one described in formula 11 is different in its chemical composition from all the others known. This product does the process of hair straightening by destroying the hydrophilic properties of the hair fiber by forming condensation products with carboxyl compounds of amino and imino groups present in the keratin molecules at a pH of below 7 which is conducive for such a process.

Formula 11

Cresol sulfonic acid	7.5%
Isopropyl naphthalene sulfonic acid	7.5%
Formaldehyde	10.0
Water	75.0

Yet another approach to hair straightening was brought about by the introduction silicones. These products mostly containing solvents are applied to the hair and given the desired shape by pressing. The silicone film which is by nature water resistant, prevents the moisture from reacting with the hair shaft thereby maintaining the dehydrated hair temporarily straightened or styled.

However, a great deal of improvement is still required in silicone films in order to enhance its water resistant properties.

7

Hair Dyes

Temporary colouring materials for hair are products used at any time to effect a physical change in the shade of the hair. These include powders, crayons, lacquers, and certain rinses. But all should be easily removable and should not have any appreciable effects on the texture or other characteristics of normal hair. Historical and cultural heritage reveals that Greeks, Assyrians, Persians, Chinese, Hindus used hair colouring materials in social life and customs.

The use of powders for changing the colour of the hair is mostly for theatrical make up, masquerades, and pageants in which it might be required for historical costumes. Otherwise it is limited to the occasional freshening of the Wigs of Judges in Courts of Law.

POWDERS

The primitive coloured powders consisted of materials for plant materials, metallic compounds, and mixtures of these two types. Later on synthetic organic dyes brought a new era in hair colouring materials. As these are all proprietary products, exact composition of the recipes are not possible. However, a broad classification of hair dyes and their formulas are given below.

Classification of Hair Colouring Agents.

1. Bleaching agents.
2. Temporary colourings.

3. Natural organic dyes.
4. Synthetic organic dyes.
5. Inorganic dyes.

BLEACHES

Roman ladies admired the golden hair of many of the captives brought from Northern Countries and tried to imitate them. Native minerals such as rock, alum, quicklime, crude soda, and wood ash, occasionally combined with old wine (or drug of wine) and water, served as favorite "Blond washes". These preparations were left on the hair for long days and the resulting reddish gold shades were retained.

Alum, ashes, borax, crude soap and soda of the earliest mixture, were still used, but always with decoctions of various plants. To quote Birch bark broom, celandine, lupine, mullein myrrh, saffron, stavesachre, turmeric, old wine, and drugs of wine etc., are a few plants.

The golden red (Venetian blond) shade of hair was immortalized. It was done by combing with a solution of soda (or rock alum, black sulfur, honey) through the hair and spread over the broad brim of a crownless hat for drying in sunlight. This treatment was also introduced in France in the 16th Century.

CHEMICAL BLEACHES

Oxygen releasing compounds were the main source of chemical agents to bleach the hair. However, for use on living human hair the chemical agents selected should be nontoxic, mild in action and free of harmful residue. Hydrogen peroxide is the most satisfactory bleaching agent for human hair. Drug store peroxide a 3-4% solution generating 10-12 volumes of oxygen is commonly used for bleaching the hair at home and 5-6% (12-20 volumes of solution) are generally used in beauty shops. Stabilized hydrogen peroxide with acetanilide or dilute acids, p-hydroxy benzoate or sodium stannate or stannic hydroxide may be used.

By applying with different proportions of peroxide and ammonia and time of contact, a wide variety of beautiful shades can be obtained.

COLOURING OF HAIR

A variety of products are used to affect the colour of the hair like powders, crayons, lacquers and certain rinses. But all of it should be easily removable. A general hair colouring powder consists of

Formula 1

Talc	20%
Starch	25
Potato meal	45
Powdered orris root	10

CRAYONS

These are sticks or blocks of colouring materials with handy shapes originally introduced for retouching of new hair in between permanent dye applications. These are made of mixtures of natural or synthetic waxes, with soap such as triethanolamine stearate, into which dyes or pigments are thoroughly incorporated (in different shades) simulating the average colours of natural hair so that they blend with other dyes that may be on it. The consistency of the finished product is such that the colour can be easily applied as the stick is rubbed directly over the hair or transferred from the stick to the hair.

Formula 2

Gum arabic	27.5%
Sodium stearate	15.0
Glycerol	15
Water	26.5
Colour	16

Procedure

Mix water and glycerol add gum arabic to half of this solution and allow it to stand. Add stearic acid to reminder and warm mixture until it is dissolved. Mix all ingredients and add colour and mill thoroughly. Run paste into moulds and dry in heat.

Formula 3

Triethanolamine	7.0%
Glyceryl monolaurate	5.5
Gum tragacanth	2.5
Stearic acid	13.5
Bees wax	50.0
Carnauba wax	13.5
Ozokerite	8.0
Colour	q.s.

Procedure

Heat the first three ingredients to 70°C. Add stearic acid and raise temperature to 75°C. Melt waxes at 75 to 80°C. Add them to other mixture and stir until well blended. Add colour and mix thoroughly. Pour into moulds at 68 to 70°C.

COLOURED RINSES

Coloured rinses or tint rinses for the hair were originally patterned on a similar product commonly used to restore or change the shade of lingerie, curtains, or other articles used in the household.

These are similar to acid rinses used as accessories to a shampoo. The base may be tartaric acid, adipic, citric, acetic and other acids have also been used alone or in combination. All dyes should be selected from the officially permitted food colours of FDA.

Table I: Certified Colours for Hair Rinses

FDA Designation	Common Name	FDA Designation	Common Name
FD&C Yellow No.1	Naphthol Yellow S	Ext D&C Red No.10	Azo Rubin Extra
FD&C Yellow No.5	Tartrazine	Ext D&C Red No.13	Croceine Scarlet
FD&C Yellow No.6			
FD&C Green No.2			
FD&C Orange No.3			

Sunset Yellow FCF			
Light Green	D&C Yellow No. 10	Metanil Yellow	SF (Yellowish)
Ext D&C Yellow No.1	Quinoline Yellow WS	FD&C Green No.3	Fast Green FC
Orange G	D&C Green No.5	Alizarine	Cyanin
D&C Orange No.4	Orange II		Green C
Ext D&C Orange No.3	Orange I	D&C Green No.7	Acid Fast Green
FD&C Red No.1	Ponceau 3R	FD&C Blue No.1	Brilliant Blue FCF
FD&C Red No.2	Amaranth	D&C Blue No.2	Indigotin (C)
FD&C Red No.4	Ponceau SX	D&C Blue No.4	Alphazurine FG
D&C Red No.22	Eosin YS	Erioglaucine	
D&C Red No.33	Acid Fuchsin D	Ext D&C Blue No.1	Methylene Blue
Fast Acid Fuchsin CB		FD&C Violet No.1	Acid Violet 6B
Naphthalene Red B		D&C Brown No.1	Resorcin Brown
Ext D&C Red No.3	Violamine R	D&C Black No.1	Naphthol Blue Black
Ext D&C Red No.6	Rose Bengale TDK	Ext D&C Black No.1	Fast Black BB
Ext D&C Red No.8	Fast Red A	Sulfo Cyanine	Black B

To produce the desired tints, dilute solutions (1 gm/litre) should be tested, singly and in mixtures, on white wool or hair and also on naturally coloured hair of shades on which the rinses might appropriately be used.

Formulas for Coloured Rinses	7	8	9	10	11
Certified Colour	Gold-Blond	Auburn	Brown	Black	Platinum
FD&C Yellow No.5	64%	—	—	—	—
FD&C Orange No.1	20	76%	—	—	—
FD&C Red No.1	3	11	—	—	—
Ext D&C Red No.1	—	—	—	22%	—
Ext D&C Red No.13	—	—	60%	—	—
D&C Violet No.1	8	13	—	—	100%
D&C Brown No.1	—	—	28	18	—
D&C Black No.1	5	—	12	60	—

These are manufactured by thoroughly mixing the acids and packed in moisture proof cellophane envelopes or sacs. Solid forms are also available like tablets, capsules, to be dissolved in measured volume of water. Surfactants are also incorporated at times for uniform diffusion.

For Grey Hair

Rinses, based on Methylene Blue, Acid Violet 6B, and water soluble Nigrosine are sold both as dry crystals and in solutions, compounded with adipic, citric or tartaric acid.

Colour Shampoos

Azo dyes have been combined with shampoos of different types. The earlier products of this kind, used a base of neutral soap in small cakes, but they were dissatisfactory. Present colour shampoos are based on synthetic surface active agents, with wetting and detergent properties and with colours ranging over 10 shades suitable for blond hair.

Wave Sets

The mucilages and gums both natural and synthetic which are used for setting the hair in finger waving or sculpture curls, can also serve as carriers for various colours.

Lacquers

These are made of Polyvinyl Pyrrolidone (PVP), Ethyl Cellulose (EC) or Shellac.

PERMANENT COLOURING

An ideal permanent colouring substance for hair should possess the following qualities.

1. Must not be injurious to general health.
2. Must dye the hair but not the skin.
3. Should not have any ill effect on the structure of the hair.
4. Must not take long time for the production of the effect.
5. Must not have any irritating effect on the skin.
6. Must produce natural shades, lasting reasonably.
7. Must be compatible with other treatments such as permanent waving etc.

The permanent dyes are of three categories.

1. Natural organic.
2. Synthetic organic.
3. Metallic preparations.

Natural Organic Dyes

The earliest dyes were natural organic substances obtained from plants, used either as in fusions, decoctions or packs. Many of the decoctions of woods are still utilized for the dyeing and blending. But a majority of the compositions contain henna or chamomile which are still being persisted with even in modern professional and amateur practices.

Henna and Henna Mixtures

Henna is a naturally occuring shrub in Arabia, Tunisia, Persia, India and other tropical countries. The active dyeing principle of Henna is lawsone. Commercial henna rinses are extracts of henna, (with boiled water) acidified with mild organic acids like adipic, citric acids.

Henna Pack: Freshly ground paste of henna leaves with 2% sodium lauryl sulphate or soapnut powder.

Henna Shampoo

Attempts were made to make a coloured shampoo, but were found to be ineffective. However, 5% henna extract mixed with sodium or triethanolamine sulfate and 5% of boricor adipic acid could give a satisfactory shampoo.

Henna Mixtures

To modify the reddish shades of henna, dried and powdered leaves of Indigofera (called reng) are either mixed with henna, or applied on the hair alternatively in a paste form. A dark brown to black shade can be obtained by changing the proportions of both the leaves.

Chamomile

In Western Europe, Great Britain and USA two varieties of Anthemis nobilis (Roman Chamomile) and Maticarra Chamomile (German or Hungarian) are used in extract form. The colouring principle is apigenin 4,5,7 trihydroxy flavone. A combination of Henna and Chamomile can also be used as a rinse, a pack or a shampoo.

Commercial Rinse

Powdered (Roman) Chamomile	40.0%
Oil of Chamomile	0.2
Alcohol 20% to make	100.0

Extract can be obtained by percolation.

Chamomile Pack

Chamomile florates, powdered mixed with Kaolinor fuller's earth and made into a thin paste with boiling water.

Chamomile Shampoo

Marketed either as powder or as a liquid.

German Chamomile (Powdered)	10%
Mild organic acid	5
Sodium lauryl sulphate	85
Oil of Chamomile	q.s. for perfume

Liquid Shampoo

Chamomile infusion	10%
Sodium lauryl sulphate	30-40%
Water	60-50%

Chamomile-Henna mixtures

Chamomile	75
Henna	25

Wood Extracts

Decoctions of woody fibres bark or nuts from several trees are also used as dyes.

Catechu

Acacia catechu is brown to black. The active principle is catechin (Catechol). Catechu and tannic acid in combination produce all shades from blond to black and act as mordants as well.

Fustic

The wood of chlorophora (Morus), Machura tinctoria in which the active principle is morin. This produces yellowish to brown shades under different conditions.

Log Wood: (Black wood, Compectic wood) : The heartwood of

haematoxylin, gives an active principle haematoxylin which when oxidised is responsible for imparting black colour and nullifying red shades of the hair.

Nutgalls

These are pathological excrescences on leaves and twigs or other tissues of white oak tree (Quercus infectoria) caused by invasion of bacteria, certain worms, or insects. These little nuts which are especially rich in tannin and gallic acid, serve as a source of pyrogallol which in turn is the source of rastiks, (hair dye).

Quercitron

The inner bark of another species of oak tree Quercus infectoria, gives an active principle quercitin. This bark is usually combined with rustic and logwood to produce dark brown shades.

Wallnut

The leaves of walnut trees Juglans cincrea. J.nigra and J.regia were used as a source for imparting a brown shade. But as the colour is not long lasting, the ground shells are used in mixture with other colour producing substances.

Mixed Wood Dyes

Henna is occasionally mixed with some of the wood extracts: for example nutgalls or wallnut shells to modify the natural orange red shade of henna alone. But unfortunately it is not long lasting.

Miscellaneous Plant Products

Among other plant sources of hair colouring, rhubarb and sage. The former has occasionally been used in combination with henna, black tea, and chamomile. Its active principle is chrysophanol, which gives blond shades in alkaline media. Because of its instability it could not be promoted commercially.

Natural Dyes with Mordants

Solutions of metallic salts act as mordants- usually copper and iron salts. Generally Rastik (Turkish and Persian name for hair dye) is prepared by roasting nutgalls with little oil, copper & iron fillings and then ground to a powder, made a paste with water and when applied to hair for long hours gives a black shade.

SYNTHETIC DYES

Most of the synthetic organic dyes contain pyrogallol and also the azo and other dyes popularly known as coloured rinses. These are all oxidizable dyes. The first dye that was attempted on human beings was pyrogallol. Many of the natural substances contain pyrogallic acid oxidised from pyrogallol. Pyrogallol solution in alkaline pH gradually develops black shade/colour. Pyrogallol with plant derivatives can be formulated into effective hair dye packs.

Amino dyes

The first dye was phenylenediamine, which is effective in 1–3% of alkaline solution followed by oxidation with an oxidising solution. This became popular due to its lustrous natural black colour in contrast to other dull colours.

Many proprietary products were available commercially. After the introduction of these aminodyes like compound henna etc., p-amino phenylamine dyes were introduced into the market. The next development was towards shampoo vehicles for shampoo imparting tints.

Formula 1

Oleic/Coconut fatty acid	5-25%
Propylene glycol	5-15%
Ammonium hydroxide	q.s. to neutralise fatty acid
Water q.s. to make	100

Formula 2

Oleic acid	15-30%
Powdered castile soap	2-6
Ammonium hydroxide	q.s. to neutralise fatty acid
Water q.s. to make	100%

Oxidation of dyes developed dermatitis. Hence steps to protect the users of these products were initiated towards improving both compositions and methods of application. As a result of which dyes developed subsequently were better than the earlier ones. Amino dyes were the so called ideal dyes but for a few reported cases of allergy.

Forms of Oxidation Dyes

The amino or oxidation dyes have predominently been products for application by experienced or skilled hair dressers. The materials for a single application are packaged in two units. One containing the dye and the other the developer. The developer can be in the form of a liquid or urea peroxide tablets.

Clear Liquids

A preliminary treatment with hydrogen peroxide or shampooing followed by liquid dye treatment. The shade of the colour/dye is monitored by this method.

Shampoo Shades

These are dyes with shampoo consisting of ammonium salts or ethanolamine salts of various fatty acids and a wetting agent.

Formula A

Oleic/Coconut fatty acid	15-25%
Propylene glycol	5-15%
Ammonium hydroxide	q.s. to neutralise fatty acid
Water q.s. to make	100%

Formula B

Oleic acid	15-30%
Powdered castile soap	2-6

Modified Shampoos

Solvent for intermediates

Ammonium hydroxide 26°C	10%
Isopropanol	25%
Perfume	0.5%

Base Solvent for intermediates

Ammonium hydroxide	10%
Isopropanol	25
Perfume	0.5
Base:	
Oleic acid	35.0%
Polyoxysorbitan monooleate	10.0
Nonionic surfactant	3.50
Water soluble lanolin	1.75
Lecithin	1.25
Chelating agent	0.25
Water to make	100.00

Procedure

The dye compounds can be mixed or added to the above solvent mixtures. This serves as an emulsifier. The pH of the final mixture should be adjusted between 9.0 and 9.2.

In compounding a shampoo tint, the viscosity of both the dye-base and the mixture of base with developer must be taken into consideration. The base must flow freely, yet viscous enough to adhere to the hair.

Colour Shampoos

These are intended to impart merely a temporarary tinge of colour to hair of any shade. However, colour shampoos can be prepared with a few shades like, gold, red brown or silver dye.

E.g. of modified colour shampoo:

Ammonium lauryl sulfate	25-35%
Fatty acid alkanolamide	2-6
Water q.s. to make	100%

Cream dyes

These were introduced in Germany in 1950's. Basically these are like cream shampoos. The product/dye is packed in tightly sealed tubes from which a measured amount is squeezed out and mixed with hydrogen peroxide. These are meant for use in beauty shops.

Black Dye Combinations

These were developed as a combination of strong hydrogen peroxide controlled by an enzyme catalyst.

Miscellaneous

Amino dyes in the form of powders and capsules. There are also cakes from which the colour was supposed to develop directly on hair previously treated with hydrogen peroxide.

Metallic Dyes

A lead comb dipped in vinegar serves as a dyer. Dilute solutions of lead acetate or nitrate, sulphur, glycerol, rose water are in the composition.

Formula

Lead acetate	1.5%
Precipitated sulphur	0.2
Glycerol	1.5
Distilled water or rose water to make	100.00

Formula A

Lead acetate	0.6%
Sodium thiosulfate	1.8
Glycerol	0.5
Ethyl alcohol	6.0
Water to make	100.00

Formula B

Formula A	25%
Sodium thiosulfate	75

Procedure

Add both the ingredients and make up to 1 liter at 95°C.

Silver Dyes

All silver salts darken on exposure, Silver dyes come in two fold packing. One containing silver nitrate and the other a developer. The principle behind this is all silver salts darken on exposure to light and silver combines readily with protein forming a brown stain.

The silver dyes are dispersed in two bottles one containing silver nitrate, the other containing a developer, with sodium, potassium, ammonium hydro sulfides or pyrogallol. With sulfides, the silver sulfide (Black) is deposited directly on the hair and with the latter i.e., pyrogallol the silver salt acts as a mordant. Later varying amounts of ammonia were added to the silver solution thereby producing a fair grade of shades. After shampooing the hair, the dye is applied and washed with a restorer like ammonia or hypo silver dyes which are more satisfactory than lead dyes.

Part A:	Black	Dark brown
Pyrogallol	5%	3%
Distilled water	95%	

Part B:

Silver nitrate	15%
Ammonia water	10%
Distilled water to make	100

Procedure

Silver nitrate and ammonia water are mixed till a clear solution is obtained and stored in a dark place. Silver nitrate may also be prepared as pomade or preferably as a water miscible cream.

As there are two different solutions, the consumer may make an incomplete application. Dyes of this type therefore are not so satisfactory but are nevertheless popular.

Copper Dyes

Copper salts and potassium ferrocyanide and developers like ammonia, hydrosulfides, or pyrogallol were also suggested and used in saloons. Copper and iron were combined in one bottle with pyrogallol and sodium sulfite. These types of products start to darken on the hair on being exposed to the air. These products should be stored in a dark place.

Formula for Copper Dyes

Copper chloride	2.0%
Pyrogallol	2
Nitric acid	0.5
Distilled water	100

Miscellaneous

Salts of bismuth (chocolate brown) iron, nickel, cobalt, manganese, cadmium with a developer like hydro sulfides or pyrogallol are available but are not very popular.

Compound Henna

Henna with indigo, logwood, or any other natural colouring material derived from plants is compound henna and is sold as vegetable dyepack. Henna is a carrier of all shades.

Advantages and disadvantages of Metallic dyes

Advantages

These are basically harmless.

Disadvantages

1. In general, these dyes leave different shades on the hair.
2. The texture of the hair is likely to become rough. Regulatory affairs and clinical evaluations are not available satisfactorily.

p-Phenylamine dyes were discouraged due to complaints from customers and the sale was controlled by FDA and a preliminary test for Dermatitis was imposed. Similarly clinical testing is obligatory before launching the product into the market.

Dye Removers

Most of the dye removers contain decolorisers like hydrogen peroxide or hydro chloride. However, there is no single safe, quick-removal process for all old dyes. A cosmetologist or an experienced hair dresser only can do this. Soaking in oils or mixture of oils is the safest removing of hair dye. Sulfonated castor oil with salicylic acid is another example of a safe henna dye remover.

Sodium hydro sulfide $Na_2S_2O_4$ in 2.5% solution is also in practice.

Future Scope

Hair dyes and dye removers, preferably natural are desirable. Although henna indigo mixture is still the most recommended, some newer plants need to be discovered and brought into formulations. Scope for research in this field is very vast.

8

Cosmetics
for Nails

THE COSMETIC INDUSTRY has a large and profitable share of prod ucts that take care of finger nails and more recently toe nails. To counter the grime and rough treatment women resort to nail cream which is used every night and get a manicure treatment weekly or biweekly. Finger nail enamel is used even more often. Pretty finger nails are a prized possession for both genders.

Nails are transparent and protective coverings to the fingers and toes. The tender skin underneath the nails gives them a natural pink colouring. Rosy nails suggest good health. Infact, doctors have used this indication to diagnose circulation of blood. Pale nails indicate poor health. Deformation of nails by biting is seen as a symptom of nervousness. The danger of long nails and as a result the possibility of broken nails may result in infection, particularly felons which may further result in the loss of the entire finger. In any case proper care of nails is a wise proposition.

Anatomy of Nails

The nails grow out of the cuticle or horny layer of the skin in the cells of the nail matrix. These cells consist of granular layers which have the capacity to grow. The matrix is located within the half moon or white arc at the base of the nail. This half moon, called lunula, is not visible on all fingers and toes. The nail does not get a direct blood supply but is connected to blood vessels through the nail bed. Nail growth varies from person to person from season to season. Like unhealthy hair is characterized by poor growth, abnormal condition of the nail reflects ill health in other directions. There is no application that stimulates nail growth. An injury to the matrix results in a permanent injury to the nail. A healthy nail grows at the rate of 1/32-1/16 inch per week. Proper care of the nail by use of friction which increases blood circulation and emollient creams to keep the cuticle surrounding them clean helps in maintaining healthy nails.

Manicure Preparations

The care of nails is called manicuring. Manicure preparations have become an important part of Cosmetic industry and the sale of these products is ever raising. These preparations include:

1. Nail Polish or Enamel
2. Polish Remover
3. Powder Polish,
4. Paste Polish
5. Nail Cream
6. Nail Bleach
7. Cuticle Remover
8. Cuticle Softener
9. Nail Tint

The most popular amongst these preparations is nail enamel. It was first popularized as a colourless transparent coating. Later it was slightly tinted. Presently this coating is sold in limited quantities. However, the demand for an opaque lacquer in numerous shades which lends the nails a smooth, flashy, flexible coating and covers any blemishes is high. Red and shades of red are popular.

FORMULAS

The manufacturing of nail enamel involves the risk of fire hazards due to the handling of inflammable materials that are used in it. The majority of cosmetic houses prefer readymade enamel which is suitably coloured and perfumed and promptly bottled. In certain large cities the manufacture of nail lacquer is prohibited by the fire department. In case of requirement of moderate quantities of nail enamel, it is advisable for the manufacturer to purchase it rather than make his own because it is economical as well as convenient.

The composition of a majority of nail polishes have a base like triacetyl cellulose of nitrocellulose, a resin, a plasticizer, a solvent, a dye or pigment and a suitable perfume.

The use of nitro cellulose alone would result in the cracking of the film of enamel not spreading evenly or adhering properly to the nail. It is at this juncture that resins, plasticizers and solvents come into play. The more common natural resins used are shellac, benzoin, dammar, sandarac, and ester gums. Plastic resins like polyvinyl and polystyrene also yield satisfactory results. They help in the formation and adherence thereof to the nail. The plasticizers generally consist of high boiling liquids like the high boiling glycol esters, tricresyl phosphate, triphenyl phosphate, dibutyl pthalate and castor oil. Camphor a solid is also used. Their function is to preserve the flexibility and adhesive power of the film. The solvents thin out and disperse the finger nail enamels and should be of such a character that the lacquer sets in 60 to 90 seconds after application. The solvents used in combination or alone are acetone, alcohol, amyl acetate, ethyl acetate, amyl format and ethyl propionate. The selection of a solvent is very important in preventing the phenomenon called "Blushing" which is a clouding of the polish by cooling the surface of the nail below the atmosphere's dew point. On the other hand two rapid drying is undesired lest an uneven film will result. The selection and blending of the pigments is most important. About a dozen basic colours make up the colour range of finger nail enamels. Of course all of them must comply with the terms of FDA, should disperse well, be resistant to light and alkali found in kitchen cleaning and dishwashing, be non-staining and must produce good gloss. Only colours that are compatible must be used to avoid separation. Insoluble lake colours with a small percentage (5%) of titanium dioxide are preferred. The mineral colours like, iron oxide, the ochres, siennas, lamp black and ultramarine are useful for blending purposes. Purchase of nail polish from a brand house is recommended rather than going for an untried or unknown product.

Perfume is really not a must in nail enamels but adds to its finer qualities. It is used mainly to suppress the unpleasant odour of the solvents used. The recommended dosage is about 1% and it should be carefully selected so that it does not affect in any way the finished enamel.

There are certain basic requirements for evolving a good nail enamel or lacquer. The glossing materials, resins, plasticizers, solvents, pigments, and perfumes must be judiciously balanced to fulfill the following conditions.

1. There must be a proper flow to produce a glossy, even film.
2. The finished film should not be sticky.
3. Must possess proper adhering qualities.
4. The film must not crack or peel off after application.
5. Must not develop clouding or blush.
6. Must be resistant to water as well as soapy, slightly alkaline water.
7. Must be light proof and not fade away.
8. The film must not dry out the fingernails.
9. The colours must be non-staining.
10. The shades must be uniform and popular.

To formulate one such polish requires skill and knowledge. With the arrival of new ingredients into the market, improvements are being constantly made.

Here are a few formulas which describe the approximate composition of nail enamels.

Nail Enamel

Formula 1

Nitro cellulose (I Second Vis)	10.0%
Ethyl acetate	50.0
Butyl acetate	20.0
Diethyl phthalate	15.0
Camphor	4.5
Colour (dye)	0.5

Dissolve the nitro cellulose and the camphor in the ethyl acetate. Add the rest of the ingredients. Perfume and colour as desired.

Formula 2

Celluloid clear	25.0%
Acetone	50.0
Amyl acetate	25.0
Oil Pink	0.2
Rhodamine B	0.01

Formula 3

Nitro cellulose dry	10%
Anhydrous methyl alcohol	75
Ethylene glycol monomethyl ether	15

Formula 4

Glycerin	0.5%
Purified sulfuric ether	4.65
Zinc oxide	15.0
Acetone	19.0
Amyl acetate	18.0
Butyl alcohol	22.0
Pure celluloid	20.0
Olive oil	0.5

Ultramarine	0.15
Lavender oil	0.2

Nail polish removers consist of a single solvent like acetone or combinations of solvents like:

Nail Polish Removers

Formula 5

Butyl stearate	60.0%
Ethyl acetate	40.0

Formula 6

Butyl stearate	35.0%
Isopropyl alcohol	65.0

Formula 7

Castor oil	10.0%
Dibutyl phthalate	40.0
Ethyl acetate	50.0

Formula 8

Glycol methyl ether	97.0%
Peanut oil	2.5
Perfume	0.5

Formula 9

Glycol ethyl ether	46.0%
Ethyl acetate	15.0
Paraffin	5.5
Beeswax	6.5
Stearic acid	20.0
Triethanolamine	6.0
Perfume	1.0

Before the advent of fingernail enamel, polishing powders, and pastes enjoyed as much popularity as the enamel enjoys today. Even today they find considerable use among certain classes.

Nail Polish Powders

Formula 10

Tin oxide	90.0%
Titanium dioxide	8.0
Oleic acid	2.0

Formula 11

Tin oxide	83.0%
Carnauba wax	4.5
Fuller's earth	6.0
Oleic acid	4.0
Tincture of benzoin	2.0
Benzyl benzoate	0.5

Formula 12

Tin oxide	70.0%
Talc	20.0
Rice starch	10.0

Nail Polish Pastes

Formula 13

Tin oxide	25.0%
Carnauba wax	15.0
White beeswax	5.0
Petrolatum, short fibre, soft	50.0
Titanium dioxide	5.0

Formula 14

Rosin	8.0%

Spermaceti	5.0
Carnauba wax	5.0
Tin oxide	30.0
Zinc oxide	10.0
Petrolatum, short fibre, soft	30.0
Kieselguhr purified	-12.0

Melt the rosin, spermaceti, carnauba wax and petrolatum together. Sift in the dry materials while mixing as before. Add a suitable perfume.

Creams for direct application to the nails and for softening the cuticle enjoy considerable popularity particularly among "manicurists".

Here are some illustrative formulas of Nail cream.

Nail Cream

Formula 15

Petrolatum, short fibre	24.0%
Cholesterol in absorption base	20.0
Cetyl alcohol	5.0
Cocoa butter	5.0
White beeswax	12.0
Borax	0.5
Water	23.0
Tincture of benzoin	10.0
Perfume	0.4
Preservative	0.1

Heat the water and dissolve the borax in it. Melt the petrolatum, absorption base, cetyl alcohol, cocoa butter, and beeswax. Add the borax solution and mix thoroughly. Add the preservative and perfume, dissolved in the tincture of benzoin.

Cuticle Softener

Formula 16

Cholesterol in absorption base	25.0%
Petrolatum	17.0
Beeswax	10.0
Sulfonated castor oil	24.0
Sodium lauryl sulfate	4.0
Trisodium phosphate	2.0
Water	17.5
Perfume (alkali stable)	0.5

Melt the first four ingredients together. Dissolve the phosphate in water, add the sodium lauryl sulfate and heat to 80°C. With rapid stirring add the melted fats. Mix until cool and add the perfume.

Cuticle Cream

Formula 17

Lanolin absorption base	25.0%
Lanolin	10.0
Mineral oil	19.0
Water	45.0
Perfume	1.0

Formula 18

Anhydrous lanolin	10%
Paraffin wax refined	10
Beeswax	25
White rose oil	5
Petrolatum Superla white	5
Distilled water	25
Boric acid crystals	1
Alkanet colour (1:10 in mineral oil)	7
Perfume	12

In manicuring it is desirable to have the protruding tips under the nails appear white. To atain this condition nail white, also a nail bleach, is used. These nail whites consist of a white pigment dispersed through a fatty base. These formulas have been used.

Nail Bleach (Nail White)

Formula 19

Titanium dioxide	20.0%
Talc	20.0
Zinc peroxide, purified	7.5
Petrolatum	26.0
Mineral oil	26.5

Mix the petrolatum and mineral oil. Slowly add the dry materials. (May sweat due to light oil content).

Nail White

Formula 20

Stearic acid	22.0%
Mineral oil	7.5
Triethanolamine	2.0
Titanium dioxide	7.5
Glycerin	7.0
Water	53.0
Perfume	1.0

A nail bleach may be made as follows:

Liquid Nail Bleach (Nail White)

Formula 21

Oxalic acid	0.25%
Citric acid	10.25
Rose water	89.5

Dissolve the acids in the rose water. Straight hydrogen peroxide is also used as a nail bleach.

Due to the universal smoking of cigarettes many finger, particularly under the nails, become stained with the volatile portion of the cigarette tobacco or paper. The products to remove these are named nicotine removers.

Nicotine Removers

Formula 22

Beeswax	10.0%
Paraffin	5.0
Mineral oil	46.0
Pumice	8.0
Borax	0.5
Water	30.0
Perfume	0.5

Formula 23

Glacial acetic acid	40%
Aluminium sulfate	32
Fine pumice	8
Water	18
Perfume	2
Mould while moist	

The cuticle around the nail has a tendency to grow over the nail in a thin irregular layer. This is usually removed by softening it with a cuticle remover and then pushing the softened layer down and off the nail with an orange stick. Cuticle removers consist of caustic alkalies and must be handled cautiously. Two typical formulas are given:

Formula 24

Potassium hydroxide	3.0%
Glycerin	8.0

| Rose water | 50.0 |
| Water | 39.0 |

Formula 25

Sodium hydroxide	5%
Water	44
Alcohol S.D.	40
Propylene glycol	10
Perfume	1

9

Antiperspirants and Deodorants

COSMETIC DEODORANTS these days are available in a variety of forms in practically every store. These products remove or decrease the malodour of perspiration, prevent its development, or do both. A survey shows that lately these products are becoming increasingly popular and have flooded the market. There is at least one advertisement of deodorants during every television programme. It shows the growing public awareness of the unpleasantness of perspiration odour, and the will to do something about it.

The odour of perspiration varies from person to person. The causes may be several. It could be the physical condition, activity of the person, emotional state or diet. To combat this unpleasant odour it is imperative to develop some products to counteract the smell, or reduce the flow of excess perspiration.

A variety of products with astringent properties are said to react with the proteins of the skin and cause coagulation, causing swelling which in turn blocks the pores of the skin, reducing the flow of sweat.

The salts of such metals as aluminium, iron, chromium, lead, zirconium, mercury and zinc have astringent properties. But not all are used for preparing anti-perspirants, as some produce discoloration, and others possible toxic effects. Basic formate, lactate, sulfamate and alums are also found in anti-perspirant products. Acetates are generally not used because of their distinctly strong odour. Aluminium salts are generally used in concentration of 12 to 20%.

There are some more pharmacodynamic agents such as Banthine, Prantal and others, which can suppress perspiration when applied by iontophoresis. Sweating can be checked by Priscoline and Regitine, by anti-adrenergic drugs since all of these drugs, though effective anti-perspirants, are known to have some undesirable side effects, and hence are not suitable for cosmetic use.

Most of the salts which show good astringent qualities are of low pH (2.5–4.0), hence they may be irritating to the skin and corrosive to fabric. The pH of such products may be modified by the addition of zinc oxide, magnesium oxide, aluminium hydroxide or triethanolamines to reduce skin irritation. A small quality of the substance is put under an adhesive patch, on the skin for 48 hrs. A subsequent patch is applied again on the same skin area and after 10 days this will indicate the sensitizing properties of the test product. It also indicates the incidence of irritation and sensitization when a product is used repeatedly on the skin. This test is a must and is valuable for anti-perspirant products where frequent application of the same product on the same place daily is done for judging the dermatological properties.

Next comes the complex problem of fabric protection. If a readily hydrolyzed salt, like aluminum chloride is used in the product the fabric will not be effected immediately on coming in contact with the solution or cream. If the salts used are not so strongly acidic, the fabric may not be damaged. Linen, cotton and viscose rayon are very susceptible to these acidic creams. Silk wool, nylon, acetate rayon are more resistant. There is maximum damage to a fabric if it comes in contact with a cream and then is ironed without laundering.

The importance of anti-perspirants which do not cause fabric damage is shown in the large number of patents issued for various means of protecting the fabric. One of the earliest issued patents for anti-perspirants is as follows:

  1 part Al_2Cl_3 (crystals)

  3 parts distilled water

  ½ part borax

  ½ part powdered alum

The water is to dissolve the ingredients, the purpose of Al_2Cl_3 is to stop perspiration, and that of borax and alum to prevent irritation to skin that could be caused by Al_2Cl_3. Another patent for anti-perspirants is a solid, wax like stick, which absorbs and deodorizes the perspiration and at the same time is not harmful to the skin and fabric.

This product is made by a mildly astringent perspiration deodorant zinc sulfocarbonate, a cutaneous sedative, zinc oleate and wax.

Another formula which has basic aluminum formate to inhibit fabric corrosion is as follows:

Basic Aluminum formate solution	35.0 Parts
Aluminum sulfate	8.0
Crystallised ammonium alum	5.0
Boric acid powdered	3.0
Tegacid	20.0
Stearic acid	2.5
Petrolatum	2.5
Water	24.0

Several patents issued to Montenier are compositions which overcome the excessive acidity in astringents.

The other products of Montenier use certain amide and nitrile compounds – like succinimide and pyrazol to overcome the acidity of Al_2Cl_3. These are used in astringent preparations like cream, liquid cream or stick.

Wallace & Hand described a new and improved perspiration retarding or inhibiting compound which while being an active antiperspirant, is not damaging to skin or fabric.

This compound comprises of an astringent, one or more water soluble heavy-metal salts of strong acids, usually mineral acids and a normally mineral water-soluble amino compound having one or more intact reactive amino groups.

The metallic salt normally comprises a water-soluble strong acid salt of such metals like aluminum, cerium, zirconium, zinc, titanium, iron or bismuth. Salts of tin, lead and cadmium are also effective. The protective agent is an aliphatic amide soluble in the strong acid salt. Another patent to Klarmann & Gates has the following formula.

	% or in Parts
Al_2Cl_3	15.0
Tegacid	15.0
Spermaceti	3.0
Bees wax	2.0
Magnesium oxide	2.5
Water	62.5

This formula incorporates a water insoluble basic compound of a metal such as an oxide, hydroxide or carbonate of zinc, magnesium or aluminum to inhibit fabric corrosion. The hydrolysis of the astringent salt of the antiperspirant develops sufficient acidity to bring about the solution of the insoluble compound and can be used in proportions from 4 to 25%.

Another patent granted to Apperson & Rich and Son, which uses aluminum orthophosphate and aluminum pyrophosphate which not only help to be anti-corrosive to fabrics, but also serve as antiperspirant agents in the preparation. A cream preparation was made from these ingredients which is said to have reduced 32% of the tensile strength of cotton fabric. The formula is as follows:

Glyceryl monostearic acid stabilized	16%
Spermaceti	5
Aluminum sulfate (crystallized)	19
Aluminum phosphate	7
Water	53

But if in the same formula water is substituted by aluminum phosphate, there is 84% loss in tensile strength. The use of 10%

aluminium phosphate with a subsequent reduction of water shows no damaging effect on cotton fabric. A patent granted to Van Mater describes compounds of zirconium which are said to have deodorant and anti-perspirant properties. They are also said to be non irritating to the skin. The zirconium salts of the hydroxyorganic acid, which are insoluble, form water soluble salts with alkali or ammonium hydroxide and also amino or amino-hydroxy compounds.

Another German patent describes a powder made of zirconium compounds, i.e. oxide, hydroxide basic sulfate, or basic carbonate, which is soothing to the skin even in the presence of perspiration.

Thurmon describes anti-perspirants made of weakly acidic cation exchange resins together with aluminum phenol sulfonate.

These resins in the preparation absorb the amino acids in the perspiration thereby preventing decomposition of bacteria, which causes the odour. These cream preparations which have 15 to 20% cation exchange resins and 10 to 20% aluminum phenol sulfonate are gentle on the skin and cause no damage to the fabrics.

Urea is probably the most effective modifier for reduction of fabric damage. It has amino groups which are available for reaction with the acid formed by the hydrolysis of the astringent salt. At room temperatures it gives very good fabric protection, probably because, it breaks down into ammonia and neutralizes the astringent salt. However, it has some undesired properties. At high temperatures it breaks down to form ammonia and CO_2 and creams containing aluminium sulfate. Such an urea break-down will form alum crystals and CO_2 in the cream which will cause swelling and sponginess.

LIQUID ANTI-PERSPIRANTS

Liquid anti-perspirants mostly contain an aqueous or hydroalcoholic solution of an astringent salt, a small amount of humectant, a perfume, a dispersing agent for the perfume and a deodorant. They are generally applied in the form of a spray. If aluminum chloride or sulfate is used as an astringent salt, then a buffer is needed. Al_2Cl_3 is an irritation causing compound and should be used in low concentrations. Aluminum chlorohydroxide complexes can be used in a spray without a buffer. A small quantity of non-ionic emulsifier like Tween 40 will serve to spread the perfume in the solution.

Alcohol increases the rate of evaporation and prevents hydrolysis of the anti-perspirant in solution. The deodorant used must be soluble in the astrigent solution. A small amount of humectant helps in preventing deposit on the spray nozzle subsequently blocking the opening.

Here is a workable formula for an anti-perspirant spray.

Part A

Alcohol	50%
Propylene glycol	5.0
Hexachlorophene	0.1
Perfume	q.s.

Part B

Aluminum chlorohydroxide	15.0
Water	29.9

Procedure

Dissolve hexachlorophene and perfume in alcohol and propylene glycol to make a solution A. Dissolve the aluminum chlorohydroxide in the water, and add this solution slowly to A and to keep in a closed container for 48 hours at least. Then filter before filling. Quarternary ammonium salts, such as cetyl pyridinium chloride, cetyl trimethyl ammonium bromide or the thiamine disobutyl cresoxy ethyl dimethyl benzyl ammonium chloride are also suitable for use as deodorants in anti-perspirant solutions. The compatibility of the quaternary with the astringent salt must be checked before its selection. Some quaternaries are incompatible with the sulfate ion. A simple formula for anti-perspirant deodorant using a quaternary is given below.

Al_2Cl_3	18%
Thiamine 1622	0.5
Propylene glycol	5.0

Ethyl alcohol	40.0
Water	36.5
Perfume	q.s.

Procedure

Mix the water propylene glycol and alcohol in a container with an agitator. Slowly add the thiamine and aluminum chlorohydroxide. Continue stirring until clear solution is formed.

Anti-perspirants and acids may generally react, so any perfume used should be acid stable in regard to colour and odour. For this, the perfume may be kept at a slightly higher temperature (45°) for about 3 to 6 months to observe any change in colour and odour.

Perfumes may be irritating to the skin. It is imperative that they be tested on the skin for possible irritation or sensitization. Because of its flexibility, inertness and general attractiveness polyethylene glycol is used as liquid antiperspirant in spray type containers. The spray nozzles should always be checked for spray characteristics with the solutions to be packed, as the spray can be viscous. Polyethylene containers are inert to spray ingredients, they are permeable to many perfume ingredients. It is therefore important to choose a perfume for a product, to be packed in a container with the least diffusion.

Weight found that the nature of the cosmetic container plays a vital role in determining the rate of permeation. Alcoholic preparations lost less perfume than preparations in mineral oil.

For a product to be packaged in polypropylene, it is obvious that shelf life tests are important in selecting a perfume.

ANTI PERSPIRANT CREAMS

The most widely accepted antiperspirants by consumers are of the vanishing cream type. Since the product must contain 15 to 20% astringent salt, the usual type of cream made with a soap emulsifier will not be effective. An acid stable emulsifier compatible with the astringent salt should be used. Satisfactory creams can be made with acid stabilized glyceryl monostearate with or without additional emulsifier. Several anionic emulsifiers i.e., sodium lauryl sulfate, sodium cetyl sulfate, triethanolamine lauryl sulfate or an alkyl or aryl sulfonate can be used. Nonionics such as the hexitol esters of common fatty acids (Spans) and their polyoxyethylene derivatives Tweens or ethers form acid-stable emulsions.

The concentration of the emulsifier helps to determine the finer consistency of the cream. Consistency of creams made out of cationic or nonionic emulsifiers are softer than those made with anionic emulsifiers. A nonionic emulsifier could be combined with a cationic or anionic emulsifier to form a cream of desired consistency. Glyceryl or glycol esters of stearic acid or other fatty acids are used alone or blended according to desired consistency of the cream.

Other products like spermaceti or cetyl alcohol are often used with glyceryl or glycol esters. Either petrolatum or mineral oil 1 to 3% may also be used. The total fatty content of the cream is generally between 15 to 20%. Depending on the type of astringent used will depend the concentration of the fabric damage inhibitor.

With aluminum salts, 5 to 10% concentration of urea will give good fabric protection. Some of the amides too could be used in lower percentages. A number of satisfactory humectants are available for cosmetic creams. Here are 11 desirable properties of humectants.

1. High hygroscopicity.
2. Narrow humectant range i.e., minimum change in water content with change in relative humidity.
3. Desired viscosity.
4. Low viscosity index i.e., little change in viscosity with temperature and with water content.
5. Good compatibility.
6. Low volatility.
7. Low cost.
8. Low toxicity.
9. Good colour and odour.
10. Lack of corrosive action.
11. Low freezing point.

The humectants most commonly used in anti-perspirants are glycerol and propylene glycol. Of these, under equilibrius conditions glycerol is most hygroscopic and sorbitol syrup 85% solution is the least hygroscopic. Propylene glycol has the lowest viscosity and sorbitol syrup the highest.

Propylene glycol and polyethylene glycol 400 are more volatile than glycerol; sorbitol is basically non volatile. All of them are low in toxicity. The polyethylene glycols 200 to 400 are innocuous for external use. None of the above products have an objectionable odour.

Humectants are used at 3 to 10% of the formula weight. The concentrations to prevent drying of the cream will depend on the other ingredients used. In some formulations, sorbitol with a concentration of 2% can give effective moisture-retaining properties while propylene glycol is more effective in the 5 to 10% range.

The consistency of a vanishing cream generally depends on the quality of the humectant and other ingredients used. Vanishing creams having propylene glycol as humectant show more crusting on exposure to air than the cream made out of sorbitol syrup or glycerol.

Although nonionic creams lose water when exposed to air they generally remain soft to touch and show less shrinkage and splitting.

Titanium dioxide used between 0.5 to 1% and well dispersed is a good cream whitener and opacifier. If the quantity of titanium dioxide is increased the whiteness and brightness of the cream also increases, but if this cream rubs off on to clothes, is difficult to remove as titanium dioxide adheres to the fabric.

The perfume used for the antiperspirant must be compatible with the emulsion and should be acid stable. Occasionally a perfume may cause bleeding or sepration in an otherwise stable cream. Karas discovered that aromatic chemicals and essential oils more often than not shorten the life of the emulsions.

Creams filled at too high temperature will harden in the container after cooling. Those filled too cold may be too soft and bleed. The temperature most suitable for milling and filling should be determined for given formulas by laboratory test batches.

The finished product should be stable when stored at a slighty raised temperature (40°C) for at least a week and for several days at 5°C temperature. It should remain attractive and cosmetically reusable.

ANTIPERSPIRANT LOTIONS

Antiperspirant lotions do not have as big a market as antiperspirant creams. The emulsion type lotion basically has the same manufacturing procedure as that used for a cream. The finished emulsion has to be carefully tested for emulsion stability with age and change in temperature, and viscosity with age.

ANTIPERSPIRANT STICKS

Antiperspirants have also been developed in the form of sticks. There is the waxy type using zinc sulfonate and oleate in a waxy base. It is not strongly astringent.

Another type of antiperspirant stick is made by a combination of astringent chloride with a hard wax dissolved by heating in alcohol. It will form a solid gel on cooling. If a higher ester or fatty acid is added, the texture and rigidity of the product improves. E.g. 22.5gm $Al_2Cl_3 H_2O$ 12g candelilla wax 16g stearic acid boiled in a reflux condenser with an alcoholic menstrum consisting of isopropyl alcohol 90% by volume. When all the solid ingredients are effected into a solution and cooled, and packed in a container, it forms a solid mass of smooth salve like consistency which could be applied on the skin to deodorize perspiration.

ANTIPERSPIRANT POWDERS

Of all the antiperspirant products the antiperspirant powders are the least effective. This is probably because the quantity of powder adhering to the skin is not sufficient enough to stop the flow of perspiration. One product could be made by mixing 25 parts of an alkali metal phosphate, mixed with lycopodium (5 parts) and talc (70 parts).

TEST METHODS

There are several methods for testing the efficacity of antiperspirants. The simplest is to use the precipitate of egg albumin to measure the astringency.

DEODORANTS

There are a number of products which are deodorants and not necessarily antiperspirant. Deodorants like zinc oxide, boric and benzonic acids may be satisfactorily incorporated into powders.

Hexachlorophene is an effective deodorant in either stick or powder form. The efficacy of these products is tested by application to the axillae, and then measuring the decrease in odour and cutaneous bacteria. A powder containing 0.5% and the stick containing 0.25% of hexachlorophene reduce by 99% the bacterial population of the axillarly fossae.

Topically applied antibiotics will also have deodorizing properties. Tests where 3% of Aureomycin was used proved 80% supperession of bacterial growth in axillae. Tyrothicin and Neomycin are found to have similar properties also. But a high index of sensitization to the antibiotics makes them unsuitable for cosmetic use. With new developments of antibiotics, sensitizing may reduce and antibiotics may be good ingredients for deodorants.

Ion exchange resins were found to have properties to absorb in vitro the odorous substances in sweat.

A mixture of anion and cation resin proved more effective than when used separately. When these anion and cation resins were incorporated in 20% proportion in ointments or emulsions, the deodorization is complete.

DEODORANT POWDERS

Deodorant powders, to be effective have to be mixed and ground well, so that the ingredients are well dispersed. The active ingredient may be dissolved in a suitable solvent and distributed through the powder mix. The perfume should be well blended with the talc then incorporated into the batch. Since the active ingredients are deodorants, they may affect the fragrance, hence any perfume being used should be tested over a period of 2-3 months, samples should be checked for loss of odour. Some basic formulas are:

Formula 1

Talc	70%
Light precipitated chalk	10
Boric acid	10
Zinc oxide	9
Zinc phenolsulfonate	1
Perfume	q.s.

Formula 2

Talc	84.0%
Boric acid	3.0
Chalk	12.0
Cetyl alcohol	0.5
Hexachlorophene	0.5
Perfume	q.s.

Procedure

Dissolve hexachlorophene and cetyl alcohol in a minimum quantity of alcohol and add to the powders.

LIQUID DEODORANTS

Several quaternary ammonium compounds have been tested and found relatively nontoxic and sufficiently nonirritating for use in cosmetic preparations. Some of these like diisobutyl phenoxy (or cresoxy) ethyl dimethyl benzyl ammonium chloride have been used in concentrations as high as 2%. Quaternaries are particularly suitable for aqueous deodorant preparations as they stick to the skin and are not easily washed out by perspiration. Hence the antibacterial and deodorant spray can be prepared, without toxic antiperspirant properties by dissolving 0.5 to 2% of quaternary in water or 5% denatured alcohol of cosmetic grade. Perfume and humectant may be added if desired instead of water or alcohol in this type of product.

The water-soluble derivatives of chlorophyll potassium, copper chlorophyllin, sodium derivatives of chlorophyll, sodium-magnesium chlorophyllin have satisfactory deo-qualities. Killian found that water-soluble chlorophyllins in concentrations of 0.05% deodorized malodorous stale perspiration. But, with chlorophyllin, or any other compound the efficacy is relative to the concentrations used. Chlorophyll derivatives can be used in combination with quaternaries too for the cream or deodorant solution.

If one of the chlorinated phenol derivatives (hexacholophene or bithionol) is used in liquid it must be dissolved in propylene glycol and alcohol as iron causes discoloration with these as with phenol derivatives. It is therefore absolutely necessary to eliminate iron contamination from other ingredients or from equipment when they are used.

These deodorants when mixed in 0.25% concentrations are effective antibacterial products.

DEO CREAMS

For any chlorinated phenol deodorants used in a cream, it is important to select an emulsifier, which is compatible.

Many non-ionic emulsifiers suppress the bacteriostatic activity of these products. Potassium stearate-stearic acid vanishing cream serves as a good base for this product.

Formula 3

Part A

Hexachlorophene	0.5%
Glyceryl monostearate	10.0
Stearic acid	4.0
Cetyl alcohol	2.0
Isopropyl myristate	4.0

Part B

Potassium hydroxide	1.0
Water	66.5
Propylene glycol	12.0
Perfume	q.s.

A deodorant cream of heavier body serves to absorb malodorous materials from the skin surface, at the same time prevents formation of bad odour, by its antibacterial ingredients.

Formula 4

Part A

Arlacel C	4.0%
Ceresin	6.0
White petrolatum	8.5
Mineral oil	20.0
Lanolin	4.5

Part B

Magenisium sulfate	0.15
Water	21.85
Zinc oxide	15.00
Zinc stearate	10.00
Aluminum phenolsulfonate	10.00
Perfume	q.s.

Water-in-oil type emulsions are satisfactory deodorant creams. Since they can be rubbed into the skin, and yet not make the skin feel greasy.

DEODORANT STICKS

Deodorant sticks with no antiperspirant have also been found in stick form. O'Neil had made a preparation of 25% zinc oxide, 25% boric acid, 33% spermaceti, 16.75% petrolatum, and 0.25% petroleum oil moulded in a stick form which rubs off as a thin film when applied on the skin.

Another increasingly popular type of deodorant stick is made with 5% to 10% sodium stearate, or with hard soaps 2 to 5% of humectant, the necessary alcohol and perfume.

The gel may be formed by dissolving soap in warm alcohol or saponification of an alcoholic solution of stearic acid with sodium hydroxide.

The solidifying temperature will vary with the concentration of soap. This then is poured hot into moulds.

Formula 5

Sodium stearate	8.00%
Sorbitol syrup	5.00
Hyamine 1622	0.25
Water	8.75
Ethyl alcohol SD 40	75.00
Perfume	3.00

Formula 6

Hard soap	8.00
Isopropyl myristate	10.00
Glycerol	3.00
Cetyl trimethyl ammonium bromide	0.25
Ethyl alcohol SD 40	75.75
Perfume	3.00

Formula 7

Amerchol –L –101	10.0 %
Glyceryl monostearate	13.5 %
Spermaceti	1.5 %
Tween 60	8.5 %
Isopropyl palmitate	3.0 %
Water	63.0 %
Perfume	q.s %
Hexachlorophene	0.5 %

Very few products have these ideal requirements. However, with new astringents, varied emulsifiers and attractive packaging, it is inevitable that many new products find their way into the market.

LABELING OF ANTIPERSPIRANTS AND DEODORANTS

The Federal Food, Drug and Cosmetic Act includes certain provisions which apply to the labeling of preparations for which claims are made to antiperspirant action. A product falls under the label "drug" as stated by this Act, if it is capable of reducing the flow of sweat because by doing so, it is affecting the normal bodily function. The ingredients responsible for this must be mentioned on the label, along with other information required by the Act.

Deodorants, which only claim to reduce the odour of sweat and nothing else, fall under the label of cosmetics and must comply with the requirements for such products.

If a cosmetic product claims to be antiseptic, it falls under that part of the Act which says that the product has antibacterial activity required of a preparation.

According to the federal Food, Drug and Cosmetic Act, a drug or cosmetic will be called misbranded if its label is false or misleading. Eg. an antiperspirant may be claimed to "check" or "reduce" perspiration rather than "stop" it. Deodorants preparations may be said to "decrease" or "stop" temporarily the odour or sweat.

The ultimate test of deodorants, or that of any product is by its usuage by the consumer. The ideal deodorant, is probably the one which has maximum antiperspirant and deodorant efficacy, and a pleasing perfume at the same time, easy to apply and leaves no uncomfortable residue on the skin. It should not stain or corrode fabrics, shelf life should be long without changing its efficacy.

10

Sun Screens

THE PURPOSE OF SUNSCREENS is to either scatter sunlight effectively or to absorb the erythemal part of sun's radiant energy. In other words sunscreen is not to prevent the suns ultraviolet radiation from reaching the skin but to reduce its intensity so as to enable the skin to develop its own protection against exposure. For this purpose opaque powdered materials like talc, kaolin, zinc oxide, calcium carbonate, magnesium oxide and titanium dioxide applied in dry form or used in sutable carriers serve as light scattering agents.

The true sunscreens function by absorbing the erythemal ultraviolet radiation. The first such cosmetic appeared in the United States in 1928. The product contained as its active agent a combination of benzyl cinnamate with benzyl salicylate in an emulsion. An effective sunscreen must satisfy the following conditions.

1. Should be able to absorb erythemogenic radiation.
2. While performing its primary function of absorbing radiation it should not loose its chemical stability which in turn may reduce its performance.
3. It must be neither non-toxic nor irritant.
4. It must be non-volatile and must not lose its efficacy in the presence of perspiration.

Table shows the ultraviolet absorption of several oils.

White mineral oil	0%
Poppy oil	23%

Coconut oil	23%	Cotton seed oil	26%
Peanut oil	24%	Sesame oil	39%
Olive oil	23 %		

The above table reveals, that sesame oil is the best ultraviolet absorbent oil.

List of some sunburn preventive substances.

1. p-Aminobenzoic acid, its salts and its derivative like (ethyl, isobutyl, glyceryl, esters; p-dimethylaminobenzoic acid)
2. Anthranilates (i.e., o-aminobenzoates; methyl, menthyl, phenyl, benzyl, phenylethyl, linayl, terpinyl, and cyclohexenyl esters)
3. Salicylates (amyl, phenyl benzyl methyl, glyceryl and dipropyleneglycol esters)
4. Cinnamic acid derivatives (menthyl and benzyl esters; a-phenyl cinnamonitrile; butyl cinnamoyl pyruvate)
5. Dihydroxycinnamic acid derivates (unbelliferone, methyl umbelliferone, methylaceto-umbelliferone)
6. Trihydroxycinnamic acid derivatives (esucerletin, methylsesculetin, daphnetin, and the glucosides, esculin and daphnin)
7. Hydrocarbons (diphenylbutadine, stilbene)
8. Dibensalacetone and benzalacetophenone
9. Naphtholsulfonates (Sodium salts of 2-naphthol-6 disulfonic and of 2-naphthol-6, 8-disulfonic acids)
10. Dihydroxy-naphthoic acid and its salts
11. o-and p-Hydroxybiphenyldisulfonates
12. Coumarin derivatives (7-hydroxy, 7-methyl, 3-phenyl)
13. Diazoles (2-acetyl-3-bromonidazole, phenyl benzoxazole)
14. Quinine salts (bisulfate, sulfate, chloride, oleates, and tannate)
15. Quinoline derivatives
16. Uric and violuric acids
17. Tannic acid and its derivatives (e.g. hexaethylether)
18. Butyl carbityl 6-propyl piperonyl ether
19. Hydroquinone

Several formulas with different formulations are available for illustration.

Formula 1: Suntan Oil

2-Ethyl hexyl salicylate	5%
Sesame oil	40%
Mineral oil	55%
Perfume colour and antioxidant	q.s

Formula 2: Suntan Jelly

Homomethyl salicylate	5%
Ceresin (65°C)	15%
Peanut oil	80%
Perfume, colour and antioxidant	q.s

Formula 3: Suntan Cream (Fatty)

Dipropylene glycol salicylate	5%
Lanolin, deodorized	35
Sesame oil	20
Mineral oil	20
Water	20
Perfume, colour and antioxidant	q.s

Formula 4: Suntan Lotion (Alcoholic)

Glyceryl p-aminobenzoate	3%
Propylene glycol ricinoleate	10
Glycerol	10
Alcohol	65
Water	12
Perfume and colour	q.s

Formula 5: Suntan Cream (Non- Fatty)

Part A

Stearic acid	20.0%
Cetyl alcohol	0.5
Methyl anthranilate	5.0

Part B

Ammonia (26°)	1.0
Sodium hydroxide	0.4
Glycerol	10.0
Water	63.1
Perfume	q.s

Formula 6: Suntan Cream

Part A

Diethyleneglycol monostearate	2.0%
Stearic acid	1.5%
Cetyl alcohol	0.5%
Methyl anthranilate	5.0

Part B

Triethanolamine	1.0
Water	90.0
Perfume	q.s

Formula 7: Sunshade Ointment

Calamine	15.0%
Petrolatum, yellow	37.5
Lanolin	12.5
Water	35.0
Perfume and colour	q.s.

Formula 8: "QUARTERMASTER CORPS" Sunburn Cream

Glyceryl monostearate	13.00%
Lanolin	4.70%
Propylene glycol	4.70%
Titanium dioxide	2.50
Sodium lauryl sulfate	0.05
Isobutyl p-aminobenzoate	2.00
Isopropyl myristate-palmitate	20.00
Water	53.05
Colour	q.s

Formula 9: Sun Cream

Phenyl salicylate	5.00%
Ethyl aminobenzoate	2.00
Titanium dioxide	1.00
Neocalamine	1.00
Yellow ferric oxide	0.10
Coumarin	0.10
White wax	2.00
Triethanolamine	0.50
Stearyl alcohol	8.00
Stearic acid	2.00
Glycerol	10.00
Water, distilled	68.30
Perfume	q.s

Formula 10: Sunburn Preventive Preparation

Petrolatum, light amber	36.5%
Stearyl alcohol	3.5
Mineral oil	2.0

Sesame oil	15.0
Calcium stearate	10.0
Kaolin	30.0
Sunscreen, approved	3 to 8

Treatment of Sunburn

Serious sunburns are usually referred to a physician. Lesser ones but nevertheless painful may be treated by palliative preparations preferably liquid with low viscosity. A mild antiseptic may be added to prevent infections. Tannic acid solutions which was earlier used for treatment of different burns have since lost favour with the medical authorities. While oil-in-water lotion are both cooling and soothing. Oils and greases are not used because they tend to produce the effect of a cover or a blanket on the skin thereby preventing escape of heat. Further, they may interfere with the action of analgesic and antiseptic substances present in the formulation. Propylene glycol is used in these formulations as a mild antiseptic.

Few examples of palliative mixtures.

Formula 11: Cooling Lotion (Calamine Type)

Zinc oxide	15.0%
Talc	15.0
Bentonite	5.0
Propylene glycol	5.0
Water, distilled	60.0

Formula 12: Astringent Lotion, Mild

Zinc sulfocarbonate	3.0%
Propylene glycol	5.0
Camphor water	92.0

Formula 13: Cooling Emulsion

Mineral oil, light	10.0%
Lanolin	2.5
Triethanolamine oleate	5.0
Propylene glycol	2.5
Water, distilled	80.0

LABEL DECLARATION

A sunburn preparation may be both a "Cosmetic" and a "Drug".

11

Eye-makeup Products

OVER THE AGES the practice of using products for eye-makeup, in and around the area of eyes existed.

These products include

1. Eye shadow or eyeshade cream, stick, or liquid. The cream is either an emulsion or a liquefying type.
2. Eye-cream generally is a water-in-oil emulsion, and is not actually a part of make-up; but can be considered under cosmetics.
3. Eye-brow Pencil: This is either a crayon or an extruded pencil.
4. Mascara: This is made as a cake, cream or liquid. Cake is either pressed or moulded.

RAW MATERIALS

All the raw materials that are chosen are to be pure, safe, nontoxic and non-irritating for any of these formulations. Equally the finished products should also be checked for the same qualities.

Pigments: The main purpose of eye-make up is to accentuate the area of the eye with colour. Coal tar colours alone are not permitted; only inorganic pigments and natural colours are permitted by the licensing authorities. The main colouring agents are carbon black, iron and chromium oxide pigments, and carmine NF. All these colours should be insoluble in water or oil soluble and must be very pure.

Black: Either Carbon black; or Vegetable or Charcoal black or Iron oxide black.

Blue: Prussian blue, or Ultramarine blue.

Brown: Iron oxides, Sienna shade.

Yellow: Iron oxides, Ochre shade.

Red: Carmine NF (Aluminum lake of the cochineal pigment)

In order to change to lighter shaded titanium dioxide or zinc oxide are used.

Raw Materials

Petrolatum: The white short fiber variety with a melting point of 43°C is recommended. Its physical properties and its higher stability are of great value in making these products.

Lanolin: A cosmetic grade and anhydrous variety with melting point 38°C to 40°C is used in eye shadows and mascaras for its lubricating and adhering qualities.

Ceresin: A microcrystalline mixture of hydrocarbons of complex composition, ceresin is available in various grades with melting points over a wide range for eye products. The 68°C melting point grade is recommended. Its use enables the formulator to stiffen the preparation and flexibility (malleability).

Carnauba Wax: This wax is available in several grades. The yellow No.1, melting point 85°C purified and bleached is recommended, particularly for its high melting point. It forms a film, which is water repellent, and it counteracts the solubilizing action of soaps in water, which is a desirable quality in mascara formulations. Finally, carnauba wax gives a luster to the dried application.

Special Products

Although there are many varieties of eyeshades, the liquefying type of the cream is most popular. But such a cream should possess easy spreadability, and the stability of the colour on the eyelids without giving a shiny appearance to that part of the face.

A few desirable shades of eye shadow are given below.

(a) Blue: 20 parts Ultramarine, 10 parts Titanium dioxide
(b) Green: 15 parts Titanium dioxide 10 parts Chromium oxide green.
(c) Brown: 30 parts Iron oxide (Sienna shade) 5 parts Titanium dioxide.
(d) Violet: 20 parts of Ultramarine 10 parts of Titanium dioxide + a small quantity of Carmine NF.

Darker shades can be obtained by increasing the percentage of ultramarine and decreasing the percentage of titanium dioxide.

For evening glitter, a small amount of pearl powder is pressed over the tinted eye shadow. To get a "gold" effect aluminum powder with various pigments can be used.

A general formula for a liquefying cream eye shadow.

Formula 1

Petrolatum white (I.P.)	65%
Lanolin anhydrous 38-40°C	5
Ceresin white 67°C	10
Bees wax white	5
Mineral oil 65/75	15

Procedure

Mix the required amount of titanium dioxide with colours with the melted petrolatum. Then grind the mixture through a roller mill. In a separate container, melt together the other ingredients and add the ground colour paste to this molten mixture, stirring the entire mass well. Pour the melted product directly into container, with a hand filler.

However, the proportions of the ingredients can be varied at the discretion of the formulator for e.g.

Formula 2: Emulsifying Cream Eye Shadow

Lanolin anhydrous	10%
Spermaceti	13
Petrolatum short fiber	77

Formula 3: Emulsifying Cream Eye Shadow

Petrolatum jelly	75%
Cocoa butter odourless grade	8
Lanolin	7
Cetyl alcohol	3
Paraffin wax, microcrystalline	7
Pigment	q.s
Preservative	q.s

In this formulation as it is an emulsion base, the stability should be thoroughly taken care off, and any probable problems should be encountered. They should be filled cold into the containers.

Formula 4

Stearic acid (triple pressed)	16%
Triethanolamine	4
Petrolatum jelly soft fiber type	25
Anhydrous wool fat	5
Propylene glycol	5
Water	45
Colour	q.s

Procedure

Heat all the waxes put together and melt at 70°C. Heat the water and the triethanolamine to the same temperature. Add the water

phase slowly to the melted waxes and stir until room temperature
is reached. The pigments should be ground in.

The stick eye shadow is today as popular as the cream form.

Formula 5: Stick Eye Shadow

Ceresin white 67°C	26%
Hydrogenated cotton seed oil	5
Castor oil	43
Carnauba wax	4
Mineral oil 75/85	6
Titanium dioxide	8
Iron oxide ochre shade	4
Iron oxide sienna shade	4

Formula 6: Stick Eye Shadow

Ozokerite white 70 to 75°C	36%
Bees wax yellow	18
Cocoa butter	19
Absorption base	5.5
Mineral oil 75/85	9.5
Oleyl alcohol	3.0
Titanium dioxide	3.0
Zinc oxide	2.0
Ultramarine blue	4.0

LIQUID EYE SHADOW

Suspension Type

In the liquid eye shadows category there are two types in the market. The first is suspension of a pigment in a mixture of oils. In such
a product, the pigment usually settles. The label must carry the
suggestion to shake well before using. The second being the emulsion cream type. Although the percentage of pigment in the liquid

suspensions is equivalent to that of the cream type, the possibility of settlement of pigment in the latter is much less.

Formula 7

Isopropyl myristate	20.0%
Corn oil	20.0
Mineral oil 55/65	59.8
Preservative	0.2

Formula 8

Sulfated cetyl alcohol	1.8%
Sorbitan monostearate	0.4
Propylene glycol	6.5
Methyl cellulose 400 cps	1.5
Ethyl alcohol SD	10.0
Water	79.8
Pigments	q.s.

EYEBROW PENCILS

There are two types of pencils. The crayon and the extruded pencil. The crayon is similar to a cream type eye shadow. The colours used are of the same character as those used in the latter, but the black and brown pigments are used in higher percentage.

The extruded eyebrow pencil although formulated as a crayon, is packed in wooden casing.

MASCARA

Mascara is one of the most ancient toilet preparations; being in use since biblical times. Its purpose is to make the eyelashes appear longer, thereby enhancing the beauty of eyes. Mascara should not have a tendency to run. It should permit easy and smooth applications on the lashes. It should not dry out too quickly and it should not "cake" or give a bristling effect to the lashes.

The early mascaras were simply pressed cakes containing soap and pigments. A simple but typical product of this type may be made according to the following formula.

Formula 9: Soap type Cake Mascara

Carbon black	50%
Coconut oil sodium soap	25
Palm oil sodium soap	25

Procedure

Carefully sift the pigment and combine with the soap chips. Pass the mixture several times through a mill and then through a podder. Finally press into cakes.

Although these types of products are still available in the market triethanolamine oleate or stearate is finding increasing favour in the market. These triethanolamine salts enable a product of lower alkalinity, with a lesser irritation potential.

Formula 10: Stearate type Cake Mascara

Triethanolamine stearate	54.0%
Carnauba wax yellow No.1 85°C	25.0
Paraffin 45°C	12.5
Lanolin anhydrous USP	4.5
Carbon black	3.8
Propyl p-hydroxy benzonate	0.2

Procedure

Melt the waxes add the colours and mix well. Run the entire mass through a heated roller mill, remelt the ground material and pour into molds with slow stirring.

In the above formulas the black colour can be replaced by ultramarian blue, iron oxide brown (burnt Sienna) iron oxide yellow (ochre) and other combinations of these pigments to yield almost any desired shade.

Formula 11: Stearate type Cake Mascara

Glyceryl monostearate self-emulsifying	40%
Propylene glycol monostearate self-emulsifying	10
Stearic acid triple pressed	20
Bees wax white	10
Mixed triisopropanolamines	10
Pigments	10

Formula 12: Stearate type Cake Mascara

Triethanolamine stearate	54.0%
Bees wax yellow	6.2
Glyceryl monostearate	6.2
Carnauba wax yellow No.1	18.0
Lanolin anhydrous USP	7.6
Carbon black	3.0
Mineral oil 65/75	5.0

Formula 13: Cream Mascara

Part A

Stearic acid USP (triple pressed)	9.1%	11.2%
Petrolatum (43°C)	5.5	—
Mineral oil 65/75	9.1	—
Isopropyl myristate	—	7.3
Glyceryl monostearate (56°C)	—	4.5
Propylene glycol	—	9.1

Part B

Triethanolamine	2.75%	3.6%
Water	64.45	55.0
Methyl parahydroxy benzoate	—	0.2
Pigments	9.1	9.1

Procedure

Melt part A and heat to 60°C. In a separate container, heat part B to the same temperature. Add B to A, while stirring. Incorporate the pigments in the combined mix.

Cake mascara is applied with a water-wetted brush. Moisture causes emulsification of the mascara, producing a colour " pay off" on the lashes. Cream mascara is but another version of cake mascara with the hardness of the mass cut down by the water to such a degree of softness, that the product can be filled into tubes. Cream mascara is applied with a constant amount of water, whereas the application of the cake mascara depends upon the technique of the individual user. A typical cream mascara may be prepared to the above formula. Finally there is a liquid mascara. Although good preparations of this type have been made and placed on the market, liquid mascara has not won any great acceptance, may be due to lack of difference from other preparations. It also has a drawback of both water solubility and stickiness. Recently modifications have been made with resins in alcoholic solutions, in which carbon black is suspended. Castor oil is often included in the formula, for the sake of some water resistance.

Formula 14

Gum tragacanth	0.2%	—
Ethyl alcohol	8.0	84%
Water	83.6	—
Rosin 10% in ethyl alcohol	—	3
Caster oil	—	3
Carbon black	8.0	10
Methyl parahydroxy benzoate	0.2	—
Preservative	—	—

Eye Creams

These are used on the eyelashes as well as around the eyes. Their purpose is to reduce dryness by lubricating the delicate, thin-skinned eye-area.

Analysis

Chemical, toxicological, shelf and, consumer use tests must be carried out.

12

Baby Toiletries/ Products

INTRODUCTION

BABY SKIN IS TENDER, delicate and therefore needs special care and protection for its maintenance in a healthy condition. The probable damages are through infections, weather and poor hygienic conditions. In order to keep these harmful influences away, special baby care products were developed and are still being perfected. In the orient, the tender skin is mainly protected by application of oil. These oils are mainly of vegetable origin like castor oil, gingelly oil, groundnut oil etc. And in the occident oils like olive and almond are blended with mineral oil. Then came the practice of important antiseptic lotions and emulsions followed by creams. Powders and bath soaps have also found a place in the evolution of baby toiletries. In the orient instead of bath soaps, oil bath with flour of pulses scented with mild organic aromatic ingredients is in practice even today. The baby soaps are ofcourse most commonly used in maintaining the tender skin soft and free from scaling. This chapter deals with a few important baby toiletries and reserves the herbal preparations for volume II.

Recent advances in baby toiletries include baby massage oils, baby shampoos, baby soaps etc.

Baby Toiletries

Baby toiletries include powder, oil, lotion, cream, soap, shampoo, cotton swabs, soft wash clothes and brushes.

The skin of the child is thinner, less cornified and less hairy then adults. Chemical analysis shows a higher proportion of water and extra cellular fluid minerals. There is a tendency toward peeling and flaking of stratum during the first 3 weeks in newborn babies.

SKIN CARE OF THE NEW BORN

Proper care of the skin in the newborn infant is important in preventing infection. At present the consensus seems to be manipulation to reduce the danger of infection. It is recommended that no water or oil bath be given during the first 7-10 days after birth in some countries. The buttocks may be gently wiped away from the folds of the infant's skin with warm sterile mineral oil or sterile cotton or soft cloth. Each time the diaper is changed; sterile vegetable oil should be applied to the soiled or wet areas of the skin.

In the group of external antiseptic application the preparation most commonly used was ammoniated mercury ointment. Another method in the prevention and treatment of skin infection of new born is the use of various sulfonamide ointments. But recently either antibiotic ointments or simple boric acid ointment are being used depending on the situation.

Baby Oils

During 7 to 10 days in the nursery, care of the diaper area varies, tap water, sterile water, baby oil, or lotion being used for the removal of faecal soil. Skin folds are cleansed with water and applied with oil, alcohol or left alone.

These preparations are based on light mineral oil and vegetable oils (peanut, sesame, olive, cotton seed). Some oils contain β-hydroxy quinoline, chlorobutenol, hexachlorophene as the antiseptic and an antioxidant.

Antiseptic baby oils are far less irritating than ammoniated mercury hence they are being encouraged.

Baby Lotions

The oily preparation by covering the skin with a continuous impervious, multi-molecular film might interfere with important functions such as respiration and eliminating body toxins and thus predispose among other things to miliana. So the use of lotions started. An oil in water lotion appeared to be more effective and comfortable than antiseptic oils.

Cationic Lotion

Another method for the skin care of the newborn and infant is the use of mildy acidic, antiseptic, cationic lotions. These emulsions are based on the use of quaternary ammonium and pyridinium compounds in oil-in-water lotion stabilized with nonionic emulsifiers. These lotions appeared to have a beneficial effect in decreasing the rashes on the infant's skin.

Care of the Diaper Area

The hygienic aspect of the cleaning procedures used in this area is of considerable importance to the age of 2 to 3 years. Types of treatment involved in removing gross contamination of body excretes include the use of warm water, soap and water, mineral oil, vegetable oil, mildly acid detergents antiseptic mineral oil lotions and cationic lotions. Occasionally the emollient oils or lotions are applied following the use of soap or detergent.

Diaper Rash

The urinary and faecal excretions of the infant are a constant source of irritation and contamination to the adjacent areas. The diaper, fitting in close proximity to the skin, concentrates this liquid and semi liquid material in a warm, moist, and airtight area when plastic material is used. The stool and urine passed by the infant remain in intimate contact with the skin until the diaper change. The least serious reaction to wet diapers is laceration of the skin in its folds, caused by moisture and fiction resulting in painful, reddened and weeping areas scalded in appearance. This is known as intertrigo, which further gives rise to a skin disorder.

The wet diaper is a medium for bacterial growth which may produce enough ammonia from urea in the urine to cause dermatitis commonly known as diaper rash. It is characterized by reddening of the whole diaper area.

Other contributory factors in the development of diaper rash include the use of improperly laundered diapers which are rough and coarse, alkaline soap or detergent residues remaining in poorly rinsed diapers and antiseptic solutions used to rinse diapers. Occasionally diaper rash and perianal dermatitis are caused by faulty diets and gastrointestinal disorder.

It was proved that ammonia was the result of the presence of faecal bacteria, which liberates ammonia from the urea in the urine.

$$CO(NH_2)_2 + 2H_2O \rightarrow (NH_4)_2CO_3 \rightarrow 2NH_3 + H_2O + CO_2$$

Brown and co-workers stated that it is primarily due to other ammonia producing pathogenic organisms such as B.Proteus, pseudomonas pyocanea and staphylococous aureus.

It would be ideal to change the diaper immediately after it becomes wet. The diaper left wet on this region for some period of time will give ample opportunity for formation of ammonia.

This severe skin irritations can be eliminated by treatment of the diaper itself and by the procedures used in cleaning and care of the diaper area. Cookers found that diapers treated with 1 : 400 dilution of mercuric chloride in a final rinse showed a bacteriostatic action while wet with urine, preventing the growth of urea-splitting B-ammoniagenes. Further, to inhibit the urea splitting organisms cationic quaternary ammonium salts were used.

Ex. p-diisobutyl cresoxy ethoxyethyl dimethyl benzyl ammonium chloride monohydrate (Hyamine 10-X)

Catonic Ointments/Mechanism of Action

The theory involved in treating the diaper area with a nonvolatile antiseptic is that the urine picks up the antiseptic from the diaper and carries it into the skin. It is thus brought into intimate contact with the urea splitting organisms and effectively inhibits their deleterious action. These authors stated that a water-repellent ointment superimposed between the diaper and the skin obviously defeats the purpose of treating the diaper and therefore a water miscible ointment containing the antiseptic is indicated.

Ex. Water miscible ointment containing p-diisobutyl cresoxy ethoxyethyl dimethyl benzyl ammonium chloride monohydrate.

The cleaning procedure used in the care of the diaper area include soap and water, detergent emulsions, mineral and vegetable oils, antiseptic mineral oil lotions, and cationic lotions. Occasionally oils, lotions, and creams are used as emollients following the use of soaps and synthetic detergents.

Baby Oils

The most convenient method of cleaning the diaper area is by the use of light mineral oil and lotions followed by dusting with talcum powder. The oil is applied to all parts of soiled area and removed with absorbent cotton. A residual film or oil remains in the cervices and on the skin providing an emollient hydrophobic barrier against urine as well as preventing friction and chafing by serving as lubricant between the skin surface. Later on, an opinion was expressed that oils coat the skin and seal off the glands which may precipitate miliana. So the lotions came into the picture.

Baby Lotions

The increasingly popular cleaning products probably at the expense of baby oils are antiseptic nonionic, anionic and cationic lotions. The oil-in-water lotions have the advantage over the oils in providing a source of water for the water soluble soil.

Ex. oil-in-water-lotion containing mineral oil, lanolin and 1% hexachlorophene as antiseptic measure against diaper rash.

Examples for anionic oil-in-water lotions are silicon oil, the non-irritant keratolytic and the relatively nonsensitizing bactericidal action of the hexachlorophene and the emollient oils. Cationic oil-in-water lotions include p-diisobutyl cresoxy ethoxyethyl dimethyl benzyl ammonium chloride monohydrate and ethyl pyridinium chloride.

Baby lotions have been formulated to prevent and cure diaper rash, check diaper odour, soften and hydrate the skin and provide an emollent lubricating oily film in the skin folds to prevent friction and subsequent chafing. This is achieved essentially with antiseptics, mineral oil, lanolin nonionic and cationic emulsifiers and possibly cholesterol. This oil-wax component is applied to the skin as the dispersed phase initially but after evaporation of the water it coalesces to form a continuous hydrophobic oil film.

Baby Powders

This is one of the most useful product in the list of baby toiletries and unanimously recommended by all pediatricians. The excellent slip characteristics of a cosmetic grade of platelet talc serves as a lubricant where skin surfaces are in apposition, as in the groin, between the buttocks, in the neck and in the axillae. Intertrigo, the redness and macreratium that tend to occur where cutaneous surfaces are in apposition, is very common in infants, especially in the diaper area. So the powder is carefully applied to the skin folds. It also helps to accelerate evaporation of perspiration (during hot months) acts as a water repellent and serves as a lubricant to prevent chafing.

Boric acid has been used as an antiseptic and buffer in baby powders. Later on scientists like Brooke & Boggs stated that the therapeutic value of boric acid is doubtful and its antiseptic quality minimal. Johnstone and co-workers emphasised that boric acid is added to talcum powder because it is one of the most practical buffering agents for this purpose and not as an antiseptic. 5% boric acid in talc was practically unabsorbed through the intact skin of infants even if erythema were present. They pointed out that boric acid when mixed with talc and subsequently wetted yielded much higher degree of dissociation than boric acid in solution, due to the formation of calcium, and magnesium borate. In 1954 Food and Drug Administration made a study of borated powders and confined preparations with boric acid to 5% concentration which is what is used in baby talc in the market.

Another raw material used in baby powders is corn starch in place of talc. Its supporters claim that it does not dust as freely as talc and acts as an adsorbent for moisture and is beneficial to the baby's skin.

Ex. 1:1800 methyl benzethonium chloride in corn starch as a preventive measure against diaper rash and intertrigo they found that cationic-corn starch powder was more effective in curing diaper rash and intertrigo.

Baby Creams

These generally have a higher ratio of oil-wax/water phase than the lotions and hence provide a great degree of emollience and lubrication per unit weight of product. These are used after the baby's daily bath to prevent chafing of those skin surfaces which are in apposition and for their emollient effect to soften, relieve and prevent chapping of the skin exposed to cold climates and low humidity.

These generally contain zinc oxide, antiseptics, fungicides, and antibiotics, vitamins, cod liver oil, polyunsaturated fatty acids or glycerides, antipruritics, anesthetics, sun screen or insect repellents. The antiseptics currently used in baby creams and ointments include hexachlorophene and cationics, in the same concentrations as used in lotions.

CARE OF HAIR & SCALP

The baby's hair and scalp are washed with the same soap or detergent used during the daily bath. These may be one of the so called baby soaps which are usually white hard-milled "neutral" type, "super fatted" toilet soaps, or a shampoo prepared from olive oil. Another preparation for bathing and shampooing is a detergent emulsion containing an alkyl phenoxy polyoxyether sulfonate (Trition X-200), hexachlorophene, cholesterol and other lanolin alcohols buffered to pH 5.5 with lactic acid. Other detergents which are recommended for baby shampoos include an amphoteric alkyl ethanol imidazolinium sodium carboxylate (Miranol derivative) a protein polypeptide fatty acid condensation product (Maypon 4C) and the mild foaming polyglycol fatty acid esters and polyglycol ether esters of sorbitol anhydrides (Tweens). Refer formulas 37 & 38.

The critical requisites which govern the formulation of a safe baby shampoo are pH, skin and eye irritation and eye sting. To minimize this baby shampoos should be buffered to pH of lacrimal fluid (pH: 7.2 to 7.4). Jenkin & Cournkers stated that eye tissue is not greatly irritated by slight acidity is evident from the fact that a half saturated solution of boric acid (pH about 4.8) is commonly used as a collyrium. So, it is quite safe to keep the solution at pH between 4.5 and 8.5.

Formulation and Raw materials: The ingredients used in the formulation of baby toiletries are selected from those generally used in adult cosmetics and hence conform to the standard TGA, USP and NF specification.

The raw materials for baby toiletries may include emollient oils and waxes, emulsifiers humectants, antiseptics, preservatives, antioxidants talc, corn starch, zinc oxide detergents and perfume oils.

Formulas for Baby Oils

Principal ingredient of baby oils is light mineral oil with other emollients such as vegetable oils and lanolin or its soluble derivatives. Antiseptics and antioxidants and perfume oils are generally added where desired. Chlorothymol and chlorobutanol have been replaced with hexachlorophene.

The antioxidants may include propyl gallate, butylated hydroxyanisole and tocopherol.

	1	2	3	4	5
Mineral oil, light	98.9%	73.85%	49.4%	39.4%	88.9%
Vegetable oil	—	25.00	—	—	—
Lanolin	—	—	5.0	—	—
Isopropyl myristate	—	—	19.5	9.5	7.0
Lantrol	—	—	25.0	50.0	—
Lanogene	—	—	—	—	3.0
Hexachlorophene	1.0	1.0	1.0	1.0	1.0
Antioxidant	—	0.05	—	—	—
Perfume	0.1	0.1	0.1	0.1	0.1

	6	7	8	9	10
Mineral oil, light	79.85%	87.85%	93.9%	93.9%	58.9%
Isopropylan No.50	20.00	—	—	—	—
Lanosol	—	12.00	—	—	—
Acetulan	—	—	5.0	—	—
Modulan	—	—	—	5.0	—
Myvacet Type 9-40	—	—	—	—	30.0
Lanolin	—	—	—	—	10.0
Hexachlorophene	—	—	1.0	1.0	1.0
Chlorothymol	0.05	0.05	—	—	—
Perfume	0.1	0.1	0.1	0.1	0.1

Baby Lotions Formulas

A popular product for cleaning the diaper area and for applying an emollient material to the skin is in the form of a lotion or emulsion. These lotions may be formulated with anionic, nonionic or cationic emulsifiers, as well as with nonionic-anionic and nonionic-cationic combination. The emollient antiseptics and other additives may be included.

O/W Nonionic

Part A

	11	12
Lanolin	1.00%	—
Cetyl alcohol	1.00	—
Arlacel 80	2.10	—
Tween 80	4.90	—
Velvaril silicone fluid 1000	5.00	—
Arlacel 83	—	2.00%
Atlas G-1425	—	4.50
Bees wax	—	2.00
Petrolatum	—	7.50
Mineral oil, light	35.00	35.00
Propyl paraben	0.15	0.15

Part B

Methyl paraben	0.15	0.15
Water	50.60	48.6
Perfume	0.10	0.10

O/W Nonionic Types

Part A

	13	14	15
Cetyl alcohol	0.20%	0.52%	—
Tween 60	0,50	—	—
Stearic acid	1.00	0.94	—
Bees wax	2.00	—	—
Arlacel 60	5.00	—	—
Lantrol	10.00	—	—
Isopropyl palmitate	10.00	—	—
Oleic acid	—	—	3.00%
Stearyl alcohol	—	0.94	—
Lanolin	—	1.04	1.00
Tegin P	—	—	6.50
Mineral oil, light	10.00	26.00	24.00
Propyl paraben	0.15	—	0.15
Butyl paraben	—	0.01	—

Part B

	13	14	15
Methyl paraben	0.15	0.09	0.15
Glycerol	5.00	—	—
Triethanolamine	0.40	0.52	1.50
Borax	0.10	—	—
Sodium alginate	—	0.36	—
Water	55.25	69.33	63.45
Perfume	0.25	0.25	0.25

Nonionic – Anionic Type

Part A

	16	17	18
Stearic acid	2.00%	2.5%	2.5%
Isopropyl palmitate	2.00	—	—
Lanolin	5.00	—	2.00
Bees wax	8.00	—	—
Polyglycol 400 monostearate	10.00	-	—
Glyceryl monostearate pure	-	2.00	—
Amerchol L-101	—	9.00	—
Oleic acid	—	—	1.50
Hexachlorophene	—	—	1.00
Mineral oil, light	15.00	4.50	25.00
Propyl paraben	0.15	0.15	0.15

Part B

	16	17	18
Methyl paraben	0.15	0.15	0.15
Propylene glycol	5.00	4.50	—
Glycerol	—	—	3.50
Triethanolamine	1.00	1.00	1.30
Water	51.45	75.95	62.65
Perfume	0.25	0.25	0.25

Nonionic – Cationic Type

Part A

	19	20	21
Cetyl alcohol	0.50%	0.50%	—
Glyceryl monostearate (pure)	1.00	1.00	—
Isopropyl palmitate	3.00	—	—
Lanolin	1.00	1.00	2.00%
Mineral oil, light	5.00	—	2.00
Amerchol L-10	—	8.00	—
Tegin	—	—	6.00
Propyl paraben	0.15	0.15	0.15

Part B

Glycerol	3.00	3.00	—
Sorbitol	—	—	2.00
Lactic acid	0.10	0.10	—
Emcel E-60	0.25	—	—
Hyamine	—	0.15	0.15
Water	85.90	86.00	87.6
Perfume	0.10	0.10	0.10

Antiseptic baby lotion and cream can be formulated with cationic quaternary ammonium, pyridinium and morpholinium salts, as well as with the newer amino amphoteric surface active agents which have antiseptic properties at acidic pH values. The inclusion of cationics in emulsions does not indicate necessarily a complete correlation between the germicidal properties in aquous solutions and on the skin. The lotion and cream invariably contain materials which have small amounts of free fatty acid such as would be found in glyceryl monostearate and other non ionic fatty acid esters, lanolin and bees wax. The cationic compound would be inactivated by interaction with the fatty acid on a molecular basis. Therefore an excess is required which may be calculated from the acid value of the oil-wax phase.

Formulas for Baby Cream

Although lotions are used for cleaning the diaper area and as light lubricants in skin folds the creams find greatest use in areas requiring more effective skin softening and protection, particularly following the daily bath.

The creams may be formulated both as oil-in-water and water-in-oil emulsions. The former are hydrophilic and deposit a continuous semi-permeable film on the skin. If the water content is relatively high, rapid evaporation of the external aquous phase from the skin surface may be excessively cooling, Water-in-oil emulsions deposit a continuous, less permeable, hydrophobic film on the skin. The slower evaporation of water from the internal phase characterized this type as a "warm" emulsion and may be preferable for infants.

Formulas 22 to 30 may be formulated as antiseptic baby creams by the incorporation of 0.5 to 1% of a chlorinated bis-phenol (such as hexachlorophene or bithionol) provided such creams meet the standards of antiseptic activity as described above. Cationic ammonium, pyridinium and morpholinium compounds may be added at concentrations of 0.1 to 0.25% to those formulas which do not contain free fatty acids.

O/W Nonionic Types

Part A

	22	23
Lanolin	3.00%	—
Atlas G-1702	5.00	—
Atlas G-1726	5.00	—
Bees wax	5.00	5.00%
Arlacel 60	—	3.00
Tween 60	—	4.00
Hydrogenated vegetable oil	25.00	17.50
Mineral oil, light	20.00	26.00
Propyl paraben	0.15	0.15
Antixodant	0.05	0.05

Part B

	22	23
Methyl paraben	0.15	0.15
Sorbitol	—	5.00
Citric acid	—	0.10
Water	36.55	38.95
Perfume	0.10	0.10

Formulas for Creams: O/W Nonionic Types

Part A

	24	25
Lanolin	1.00%	—
Isopropyl myristate	4.30	—
Polyglycol 1000 monosterate	6.00	—
Stearic acid	7.20	—
Bees wax	—	3.00%
Spermaceti	—	3.00
Glyceryl monostearate, pure	4.50	12.00
Mineral oil, light	—	30.00
Propyl paraben	0.15	0.15

Part B

Methyl paraben	0.15	0.15
Propylene glycol	2.50	—
Glycerol	—	8.00
Water	74.10	43.6
Perfume	0.10	0.10

W/O Nonionic Types

Part A

	26	27
Atlas G-1441	1.00%	—
Atlas G-1425	5.00	4.00%
Bees wax	2.00	—
Stearic acid	15.00	—
Lanolin	1.00	1.00
Arlacel 83	—	2.00
Paraffin	—	10.00
Mineral oil	15.00	15.00
Petrolatum	—	35.00
Propyl paraben	0.15	0.15

Part B

Methyl paraben	0.15	0.15
Sorbitol	10.00	2.50
Water	50.60	30.10
Perfume	0.10	0.10

Cream: W/O Nonionic Types

Part A

	28	29	30
Glyceryl monostearate (pure)	5.00%	—	10.00%
Petrolatum	5.00	—	10.00
Mineral oil, light	15.00	25.00%	10.00
Microcrystalline wax (175°F)	10.00	5.00	—
Amerchol L 101	15.00	10.00	—
Lanolin	—	10.00	—
Bees wax	—	—	5.00
Lanolin absorption base	—	—	25.00
Propyl paraben	0.15	0.15	0.15

Part B

Methyl paraben	0.15	0.15	0.15
Water	49.60	49.60	39.60
Perfume	0.10	0.10	0.10

Formulas for Baby Powders

The composition of baby powders has an antiseptic ingredient and is not so stongly perfumed. Talc a natural hydrous magnesium silicate $H_2OMg_3(SiO_3)_4$ remains the most important constituent of baby powders. It has excellent slip characteristics and good adhesion to the skin.

The particle size and shape are very important. The best cosmetic grades of domestic and imported talc are imperative and should pass through a 325 mesh screen (44 microns).

The optimum range for a baby powder is 10 to 40 microns. Micro pulverization may be used to disperse the perfume oils uniformly throughout the powder.

Baby powders are used as lubricants in skin folds to prevent chafing, to absorb perspiration and relieve prickly heat, and to impart a clean pleasant fragrance to the baby's skin.

3 to 5% zinc and magnesium stearate were used to give velvety smoothness to the powder and later on replaced by lithium stearate which is non toxic and has a high degree of water repellency and oil absorbency.

Olive oil can also be added to talc in place of zinc stearate to improve adherence and emollience on the skin to impart greater water repellency to the powder.

2 to 5% zinc oxide also added for its special properties.

Colloidal kaolin (a hydrated aluminum silicate) is used in baby powder for its high moisture absorptive capacity.

Excellent emollient effects may be achieved by the incorporation of 0.5 to 1.5% cetyl or stearyl alcohol in baby powders.

The starches used in baby powder as the major ingredients in place of talc with 0.5% methyl benzethonium chloride as the antiseptic. They do not dust as talc and have absorptive capacities for moisture. The disadvantage with the starch is that it forms a sticky paste when moistened with water and acts as a substrate for bacteria.

A chemically modified corn starch known as Dry-Flo (142 to 145) is available as an ester which imparts water-repellent characteristics to the starch molecules.

The perfume selected must have great stability against oxidation and polymerization.

	31	32	33	34	35	36
Talc	99.75%	97.25%	97.25%	92.25%	90.25%	96.75%
Lithium stearate	—	2.50	—	2.50	2.50	—
Olive oil	—	—	2.50	—	—	—
Kaolin	—	—	—	5.00	5.00	—
Zinc oxide	—	—	—	—	2.00	3.00
Perfume	0.25	0.25	0.25	0.25	0.25	0.25

	37	38	39	40	41
Corn starch	99.70%	—	—	—	—
Rice starch	—	99.70%	—	—	—
Dry flo	—	—	99.70%	50.00%	—
Talc	—	—	—	49.70	84.75%
Calcium undecylenate	—	—	—	—	15.00
Hyamine 10-X	0.05	0.05	0.05	0.05	—
Perfume	0.25	0.25	0.25	0.25	0.25

Formulas for Baby Shampoo

Sodium lauryl ether (3.0) sulfate 70%	6.00
Cocamidopropyl betaine 30%	12.00
Polyoxyethylene (80) sorbitan monolaurate	6.00
Polyethylene glycol distearate	1.50
Water	74.50
Perfume, preservative and colour	qs

Tearless surfactant (Miranol)	40.00
Sodium lauryl ether (2.0) sulfate	5.00
Propylene glycol	5.00
Lauramide DEA	2.00
Water	48.00
Perfume, preservative	qs

13

Shampoos

IT IS BELIEVED that the word "Shampoo" has its origin in Hindustani language meaning to "Squeeze".

While before the advent of shampoo, a cake of soap was used for washing hair, today shampoo is very popularly used for the purpose by both men and women. Originally shampoos were of soaps or a mixture of soaps, whereas synthetic detergents are used in the majority of commercial shampoos.

DEFINITION OF A SHAMPOO

"A preparation of a surfactant (i.e. surface active material) in a suitable form-liquid, solid, or powder which when used under the conditions specified will remove surface grease, dirt and skin debris from the hair, shaft and scalp without affecting adversely the hair, scalp or the health of the user".

SHAMPOO TYPES AND FORMS

Shampoos are available in many types and forms and are classified according to their physical appearance, their special ingredients or properties. These forms are: liquid clear shampoo, liquid creams or cream lotion shampoos, cream paste shampoos, egg shampoos herbal shampoos, dry shampoos, liquid dry shampoos, colour shampoos, and aerosol shampoos. In addition to these 1) medicated shampoos 2) special shampoos 3) for tender scalp of babies (baby shampoo) are also available.

Liquid Shampoo

When a shampoo is based upon soap, it is frequently a potassium salt, when based on a synthetic, it is often a triethanolamine salt. Liquid shampoos are also made from sulfonated oils and are frequently promoted as oil shampoo or hot oil shampoos. These formulations may contain mineral oil which does not rinse out easily and thereby remains as a dressing on the hair. Liquid shampoo's popularity lies in its easy application, excellent foaming and easy rinsing.

Formula for Liquid Shampoo

Coconut oil	14%
Olive oil	3
Castor oil	3
Potassium hydroxide 85%	5
Glycerol	2
Ethyl alcohol	4
Sodium hexametaphosphate	1
Perfume	0.3
Water	68

Liquid Cream or Cream Lotion Shampoo

Liquid cream shampoo specializes on conditioning action whereby the hair is more lustrous, softer and easier to manage. Some contain lanolin, others dispersed egg powder and many contain ethanolamides of the higher fatty acids which are known for their conditioning action.

Formula for Liquid Shampoos

Triethanolamine lauryl sulfate (C_{10} to C_{18})	35%
Sodium alginate	2.5
Water	62.5

Creame or Creame Paste Shampoo

In recent years the synthetic detergents have been formulated into cream type products. By careful formulation an excellent cream paste shampoo can be obtained either from a soap or a synthetic detergent.

Formula for Creame or Creame Paste Shampoo

Water	81%
Calcium alginate	2
Sodium citrate	1
Triethanolamine lauryl sulfate	10
Glycerol	5
Methyl p-hydroxybenzoate	0.15
Perfume	0.85

Egg Shampoo

Originally egg shampoos were considered to be exclusively composed of egg. There has been a slight change in the concept of egg shampoo now wherein the shampoo is based upon a soap or synthetic detergent to which egg powder is added. These shampoos were recommended in hard water areas and for extremely dry hair. It is said to leave the hair with improved luster.

1–5% of emulsified or dispersed egg has a stabilizing action on the foam (improves the dirt dispersing action of the soap or synthetic detergent). However, addition of egg has an excellent sales promotion value due to psychological reasons rather than therapeutic efficacy.

Herbal Shampoo

These products have recently gained wide popularity although a few years back they enjoyed a very limited demand. These products had a market in hardwater areas. They are based on a whole class of saponins that are widely available in nature.

Commercially saponin is usually extracted with water and alcohol form its source. Such preparations are usually meant to cleanse the scalp and to reduce scaliness. They are usually compounded with rosemary and celandine .

Formula for Herbal Shampoo

Quillaia bark, (extract)	5%
Ammonium carbonate	1
Borax	1
Bay leaf oil	0.1
Water	92.9

Liquid Non-aqueous Shampoo

Non-aqueous shampoo is in liquid form and useful in conditions where the water is hard. They use organic solvents such as Benzene, Carbon tetrachloride etc., for cleaning.

But none of these solvents are safe as far as toxicity and inflammability are concerned. Further, since these are excellent lipid solvents they tend to strip the hair of natural grease or oil. A good shampoo should not remove and deprive the hair its natural oil.

RAW MATERIALS

Since shampoo formulations comprise of soap and synthetic detergents, their respective uses, difference and advantages would make an interesting review. Soaps are generally defined as salts of fatty acids. They were originally obtained by saponification of natural, animal or vegetable fats and oils with alkali. Of late the fats or oils are sulfonated or accurately sulfated without splitting the glyceride. Strictly speaking these sulfated oils are synthetic detergents.

Vegetable Oils and Soaps

The animal and vegetable fats are glycerides of fatty acids. The general practice over the years has been to formulate soap shampoos based upon mixtures of oils so as to obtain desirable proportions of fatty acids. In such mixtures the acids are balanced in such a way that the desired foaming and cleansing action are obtained. In general those oils containing shorter-chain fatty acids containing 10-12 linear carbon yield better foaming soaps.

Formula of Vegetable Oil Shampoo

Coconut oil	15%
Palm oil	5
Potassium hydroxide, 90%	3
Sodium hydroxide, 90%	1
Ethyl alcohol	7
Water	69

Formula of Triethanolamine Shampoo

Triethanolamine	5.4%
Oleic acid	5
Coconut fatty acids	4
Propylene glycol	5
Versene 100	0.4
Water	80.2

Olive Oil Shampoo

Olive oil soaps have been considered best for shampoo universally.

The soap formed is essentially sodium oleate which however does not produce copious lather. But it is an excellent conditioner because in olive oil or other castile shampoos some free olive oil is left in the shampoo which acts as a conditioner.

Coconut Oil Shampoo

In the case of coconut oil soaps, excellent lathering properties are found. They consist of sodium laurate and sodium myristate. The simplest coconut oil shampoo is illustrated in formula here under. Potassium or sodium hydroxide or both have been used as alkali for saponification of vegetable oils, or for neutralization of free acid. Recently triethanolamine soaps have become popular.

Formula of Coconut Oil Shampoo

Coconut oil	21%
Pottasium hydroxide, 85%	4
Water	54.1
Perfume	0.5
Olive oil	3
Sodium hydroxide, 95%	1.9
Ethyl alcohol	15
Ethylene diamine tetra acetic acid	0.5

NEED FOR SYNTHETICS

The water solubility or dispersibility of soaps is partly due to the carboxyl group at the end of the long chain hydrocarbons. However, solubilizing action is neutralized when heavy metal salts are formed, particularly calcium, magnesium and barium salts, Soaps in soft water give excellent results generally superior to all types of synthetics. However, in hard water conditions, the calcium and magnesium ions react with soaps forming insoluble salts, which are clouding rather than clearing agents. Synthetics or even a mixture of synthetics are no match to well formulated soft water soaps combining wetting, emulsifying and dispersing action while reducing surface and interfacial tension. Synthetics or mixture of synthetics on the other hand posses only one or two of these characteristics.

ALKYL BENZENE SULFONATES

It was found that if the COOH, group in fatty acids is replaced by SO_3H or SO_4H groups their soaps do not form insoluble calcium or magnesium salts.

Sodium alkyl benzene sulfonate

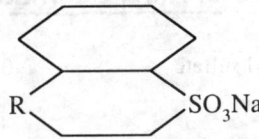

R⟨benzene ring⟩SO_3Na

Sodium alkyl naphthalene sulfonate

R⟨naphthalene ring⟩SO_3Na

In some of the earlier developments in synthetic detergents, by varying the size of alkyl group (R) detergent properties of varying action would be obtained.

Short chain alkyl group (4-8 carbon atoms) was neither a good wetting agent nor good emulsifying agent or a detergent. With 10-14 carbon atoms,

$C_{12}H_{25}$⟨benzene ring⟩SO_3Na

the product was a good detergent. Sodium keryl benzene sulfonate derived from a low cost product such as kerosene was developed which drastically cut sales of soap.

Formula of Alkyl Benzene Sulfonate Shampoo

Sodium keryl benzene sulfonate	20
Glycerol	8
Perfume	0.5
Water	71.5

PRIMARY ALKYL SULFATES

The alkyl sulfates were first developed in Germany where vegetable oils and fats were relatively scarce and detergents resistant to hard water were badly needed. A wide range of sodium alkyl sulfates were developed and the most are those of C_{12}-C_{18} series ($C_{12}H_{25}SO_4Na$ to $C_{18}H_{37}SO_4Na$). Some of the sulfates consist of a triethanolamine salt instead of a sodium salt. In some cases both these salts have been used together in a shampoo very effectively.

Formulas of Alkyl Sulphate Shampoos

Sodium lauryl sulfate	40%	–
Perfume	1	1%
Water	49	44
Triethanolamine lauryl sulfate	–	45
Ethanolamine of lauric acid	10	10

SULFONATED (SULFATED) OILS

The sulfated oils should be considered as synthetic detergents. They were first manufactured as back as 1880. One of the more important sulfonated oils is sulfated castor oil which has a probable structure of $CH_3(CH_2)_5CH(OSO_2ONa)CH_2CH{=}CH(CH_2)_7COONa$. These oils are effective in hard as well as soft water. But they tend to remove colour from the hair (natural or dye). Further, it was claimed excessive use of these oils tends to leave hair dull, wiry, and rough .

Here is an illustration of a sulfated olive oil shampoo.

Formula

Water	14%
Sulfated castor oil 75%	59.5
Sulfated olive oil 75%	19.5
Mineral oil light	3.0
Glycerol	3.5
Perfume	0.5

NONIONICS

The nonionic detergents offer a promising range of shampoos although their low yield of foam has caused limited usage. The nonionics are very water soluble or water dispersible due to the chain, $(OCH_2 \cdot CH_2)_x$, obtained by reacting ethylene oxide with a compound having an active hydrogen. Some examples of nonionics are Triton X-100, Igepal, Myrj, and the Pluronics.

Formula of Nonionic Shampoo

Sodium lauryl sulfoacetate	4%
Sodium 2-ethyl hexyl sulfoacetate	9
Sodium n-hexyl sulfoacetate	4
Sodium citrate	2
Gum karaya	1
Water	80

SHAMPOO ADDITIVES

Certain compounds which are used as shampoo additives and play an important role in enhancing the performance of a shampoo are mentioned here below.

OPACIFYING AGENTS

Here are some compounds that opacify soaps or shampoo.

Higher alcohols such as stearyl and cetyl alcohols, higher acids, such as behenic acid (22 carbons) give opacity.

The glycol mono and distearates, glyceryl and propylene glycol stearates and palmitates, are also effective opacifiers.

In some cases gaubers salt has been used, so also spermaceti.

Magnesium salts like magnesium stearate and silicate have also been used for their opacifying action.

The amides and ethanolamides of the fatty acids are used as opacifying agents as well as conditioning and foaming agents.

Finally the gums are also well known opacifiers.

CLARIFYING AGENTS

Clear shampoos share an equal amount of popularity as opaque shampoos. Butyl alcohol, isopropyl alcohol, ethyl alcohol, terpineol, pine oil, diethylene glycol, and diethyl carbitol are a clarifying agents to name a few. A sequestering agent, ethylene diamine tetra acetic acid (EDTA) is frequently used as a clarifying agent.

FINISHING AGENTS

In the ordinary shampoo vegetable oils act as conditioning agents. Recently lanolin and its derivates are being used as finishing agents although they are known to cut foaming and for their scouring action. Isopropyl myristate and butyl palmitate are particularly known and diethanoloxide of fatty acids are also used as finishing agents. Polyglycol stearate and glyceryl stearate improve finishing action. Carboxy methyl cellulose a synethtic "gum" in controlled quantities and lauryl sarcosine are also much used finishing agents.

SEQUESTERING AGENTS

The sequestering agents are added to shampoos for prevention of precipitiation of insouluble calcium or magnesium salts in hard water conditions and as well as formation of thin film of lime soap on hair when rinced with hard water.

Upto a proportion of 1% ethylene diamine tetraacetic acid (EDTA) prevents lime soap formation. The sequestering agents work well when used in sufficient quantities.

Tween 80 has been used for its cleansing action as soap and for lime soap dispersion. Tetra sodium pyrophosphate, tripolyphosphate and hexametaphosphate are a few others.

CONDITIONING AGENTS

Finishing and conditioning agents are closely connected. Humectants are usually used as conditioners. They bring moisture to the hair and reduce brittleness. Glycerol, propylene glycol, sorbitol and urea help to retain moisture and delay its evaporation. The carbo waxes and their stearates provide slip and body to the hair and the fatty acid amides slip and smoothness.

THICKENING AGENTS

Natural gums such as tragacanth gum, acacia and locust been gum were replaced by synthetic gum such as hydroxy methyl cellulose, methyl cellulose and carboxymethyl cellulose. But they too leave an undesired deposit on the hair. The alginates and the carrageenates too have had restricted use.

However polyvinyl alcohol and polyvinyl pyrrolidone are more widely used.

Diethanolamide, lactic acid or the glycol or glycol stearates are commonly used without the fear of formation of a film on the hair.

FOAM BUILDERS

Diethanolamide of lauric acid and dodecyl benzene sulfonate are good foam builders.

PRESERVATIVES

The following preservatives are preferred for preservation of shampoos.

Formaldehyde, methyl, propyl or butyl hydroxy benzoate, the alkyl anisoles, hydroxy quinoline dihydroacetic acid salts and the alkyl cresols.

ANTIBACTERIALS

Anti bacterials, for preventing bacterial skin diseases are generally added in soap type shampoos with the addition of chlorinated phenol like hexachlorphene or bithionol. Usually 2% of such a chemical is used.

EVALUATION OF SHAMPOOS

1. Foam and foam stability
2. Detergency and cleansing action
3. Surface and interfacial tensions
4. Wetting action
5. Effect in hard water
6. Conditioning action
7. Luster and softness of hair
8. Relative effectiveness
9. Irritation to eyes
10. Fragrance
11. pH
12. Soap or synthetic detergent content
13. Free salt and ash content
14. Viscosity
15. Cost and selling price

PH OF SHAMPOO

As has been pointed out by Harris, it may be difficult to obtain accurate pH reading, even with electrodes. The electrodes must be washed clean, and be free from residual traces of alkali or acid to ensure accurate readings. In any event, it is possible to prepare a shampoo at almost any pH, but there is no assurance that the pH will not drift on standing. Certainly there is no evidence that any specific range of pH gives the best shampoo. Soaps in general are more effective at a pH of 9.0 to 10.0, and synthetics may be effective at pH of 6.0 to 9.0 . It is probably true that the higher the pH, the lower the cost per shampoo. Shampoo adjusted to the isoelectic point of hair, viz., between 4.0 and 5.0 will cause better manageability.

ANTI-DANDRUFF SHAMPOO

There has also been a great deal of interest in a dandruff preventing shampoo. (Abott Laboratories has placed on the market a therapeutic shampoo which was developed based on selenium sulfide). This product, Selson, which is sold only on medical prescription, has been enjoying growing acceptance. It is said to relieve severe itching and dandruff. The following formulas are typical examples of shampoos containing the selenium and a quaternary antiseptic. However, there are undesirable side effects like greying and loss of hair.

Formulas for Anti Dandruff Shampoos

Tween 80	4%	—
Cetyl benzyl dimethyl ammonium bromide	12	—
Perfume	q.s	0.5%
Water	84	q.s 100
Arlacel 80	—	13
Glyceryl monoricinoleate	—	1
Bentonite	—	4
Selenium sulfide-bentonite , 1:1	—	5
Citric acid	—	0.4
Monosodium phosphate	—	2

ALL SYNTHETIC DETERGENT SHAMPOO BAR

Some recent formulations of a solid detergent shampoo bar:

Formula 1

Sodium lauryl sulfate(90% active)	50.00%
Ethoxylated alkylphenol	23.00

Monoethanolamide of stearic acid	23.00
Perfume	2.00
Dye (D&C Green No.5; 0.4% solution)	1.00
Polymer JR – 400 (Union Carbide)	1.00

Formula 2

Sodium lauryl sulfate (90% active)	34.0%
Sodium dodecyl benzene sulfonate	21.0
Ethoxylated alkylphenol	21.0
Monoethanolamide of stearic acid	21.0
Perfume	1.6
Dye (D & C Green No.5; 0.4% solution)	1.4

Formula 3

Monoethanolamine lauryl ether sulfate	12% or parts
Triethanolamine N-acyl sarcosinate	3.5
Alkyl dimethylammonium acetate	3.0
Quaternary copolymer of vinyl pyrrolidone	
Dialkyl lower aminoalkyl acrylate	0.25
Water, qsp	100
pH = 8.2	

The shampoo composition can be in form of a clear, opaque or pearly solution or in the form of a cream, paste, gel or aerosol. The pH of the detergent composition can range from 6.5 – 9.0, preferably from 7.5 to 8.2.

USE OF HONEY AS LUBRICATING AGENT

The formula contains a conventional detergent and honey. Optional ingredients are a foam stabilizer, a viscosity adjusting agent, perfume and a preservative. The honey is deposited on the hair by dilution of the shampoo during use and gives the hair fullness and maintains the wave and acts as a lustering agent for hair.

Formula 4

	% by weight
Fatty alcohol sulphate (detergent)	40.0
Honey	10.0
Coco-fatty acid – diethanolamide (foam booster)	3.0
Viscosity adjusting means (NaCl)	3.0
Methyl & Propyl paraben (preservatives)	0.1
Perfume	1.0
Water	42.9

Shampoo composition of basic pH

Amphoteric detergent (Tegobetaine C)	50.0 parts
Nonionic amine oxide detergent	4.0
Chamomile extract (German 25%)	3.0
Urea	2.0
Triethanolamine (85%)	1.0
Nonionic emulsifier (Tween 20)	1.5
Perfume	0.3
Water to make	100 parts total

The above shampoo has a pH of about 8.3. The mixture showed no apparent change in appearance or performance after storage of 40 days at temperatures ranging from 3 – Rt-40°c.

The addition of 2 parts of polyethylene glycol (6000) distearate, results in slight increase in conditioning effect.

Shampoo Composition of Acidic pH

Amphoteric detergent	50.0 parts
Nonionic amine oxide detergent	4.0
Chamomile extract (German 25%)	3.0
Urea	2.0
Nonionic emulsifier (Tween 20)	1.5
Perfume	0.3
Water to make	100 parts total

This resultant shampoo composition has pH value of about 5.75-Further pH adjustments may be made with small amount of an organic acid, like citric acid.

Triethanolamine Shampoo

Lauric acid	7.5
Triethanolamine	15.0
Alginic acid	3.0
Triethanolamine oleate	7.5
Coconut fatty acid chloride condensation product of collagen protein hydrolysate (Maypon 4CT)	11.5
Sorbitol 70%	10.0
Sugar	0.5
Preservatives	0.21
Sequestrant	2.0
Fragrance as needed	
Water	qs to 100

BEER SHAMPOO

Beer Concentrate

This is derived from a by-product of beer fraction produced in the conventional beer manufacturing procedure. The top layer clear beer is decanted and sent for bottling (after fermentation and addition of tannins). The bottom liquid sludge layer, often discarded in the past, is subject to seperation procedure whereby a liquid layer and a solid layer are produced. The liquid layer is further separated from the solid layer and then concentrated by evaporation of the liquid to produce a beer concentrate.

Beer Shampoo

Water	71.92%
Methyl paraben	0.25
Propyl paraben	0.05
Sodium lauryl sulphate, 60% (Standapol CS)	3.00

Sodium polyoxyethylene (1) lauryl ether sulfate	0.10
Disodium EDTA, dihydrate	0.05
Citric acid anhydrous	0.18
Methyl diethyl polyoxypropylene (8) Ammonium chloride	2.00
Preserved hydroxypropyl methyl cellulose mucilage (3.5%)	5.00
Lauric myristic (70 : 30) diethanolamide	5.00
Perfume	0.30
Beer concentrate, 50%	12.00
Formaldehyde solution, 37%	0.15

Conditioning Shampoo

A	Ammonium lauryl sulfate	22.5%
	Cocoamidopropyl betaine	5
	Lauramide DEA	2
	Panthenol	0.1
	Perfume	0.1
B	Water	69.9
	Crotein BTA	0.1
	Methylparaben	0.1
	Formaldehyde USP	0.1
	FD & C Blue No. 1, 1%	0.1

Heat A to 50°C to dissolve lauramide. Add B.

Baby Shampoo

Tearless surfactant	48.0%
Na laureth - 2 surfactant	2.0
Propylent glycol	0.1
Water	48.0
Lauramide DEA	2.0
Formaldehyde USP	0.1
	100.2

Mix in the order shown. Heat to 50°C to dissolve lauramide. pH should be 7.0.

Sodium lauryl ether (3.0) sulfate 70%	6.0%
Cocamidopropyl betaine 30%	12.0
Polyoxyethylene (80) sorbitan monolaurate	6.0
Polyethylene glycol distearate	1.5
Water	74.5
Perfume, preservative, dye, etc.	q.s.

Mix all the ingredients together by constant stirrings.

14

Bath Preparations

WHAT IS THE PURPOSE of a Bath preparation?
A good bath preparation is one that more or less fulfills the following functions.

1. To make the user feel refreshed.
2. To help to soften hard water.
3. To make the bath a pleasant and fragrant one, by means of perfume and colour.
4. To clean the body by removing dirt and odour and imparting a pleasing fragrance to it.
5. To prevent a ring from forming around the bath tub.

The bath preparations considered in this chapter are

1. Bath salts
2. Bath oils
3. Bubble bath
4. Bath powders.

Soap, although is a very commonly used bath preparation, is not being considered here because it is not regarded as a true cosmetic under the legal and other definitions of that word and its manufacture is quite different from that of the other bath preparations.

BATH SALTS

There are basically two types of bath salts: one which gives perfume and colour to the bath and the other which helps soften hard water and make cleansing easier apart from possessing the two qualities of the former.

The first type is based on rock salt crystals, which are coloured and perfumed. This salt is used because of its inert nature and is coloured (permitted) and perfumed rather easily. It does not soften water or help in cleansing either and the crystals are slow in dissolving in the water. Epsom or Glaubers salts may also be used in such preparations.

METHOD OF MANUFACTURE

The methods of manufacture is simple. The salt crystals of the desired size, a solution of the perfume oil (about 0.5 – 1 %), alcohol and the colour are all put in a ribbon powder mixer and mixed and transferred into trays to allow the alcohol to evaporate. The dry product is ready for packing.

The second type used to soften water is formulated around one of the sodium salts either sodium phosphate or sodium sesquicarbonate. Trisodium phosphate is commonly used. However, due to its high alkalinity it is generally used in conjunction with sodium sesquicarbonate or borax to buffer it. The utility of the preparation lies in its ability to somewhat soften hard water and provide at the same time cleansing action. Further, it reduces the surface tension of water thereby allowing wetting of skin.

It has a few disadvantages as well. Its high alkalinity may be of concern to people with sensitive skin. The other disadvantages occur in the manufacturing process. The perfume oil as well as colour need careful selection to prevent any possible reaction with the strong alkali on standing. The following formulas described are a few preparations of the second type.

Formulas for Bath Salts

Formulas	1	2	3
Trisodium phosphate	50 to 49%	50 to 49%	—
Sodium sesquicarbonate	49.5 to 49	—	80 to 90%
Rock salt or sodium chloride	—	49.5 to 49	—
Borax, powdered	—	—	18 to 8
Perfume	0.5 to 2	0.5 to 2	2
Colour	q.s	q.s	q.s

Procedure

The same methods described for the rock salt crystals, can be used except for first mixing uniformly two or more dry ingredients before the perfume and colour are added; otherwise the finished product might not be uniform in appearance.

Other phosphates suggested as substitutes for trisodium phosphate and would serve as better water softening as well as sequestering agents are sodium hexametaphosphate, tetra sodium pyrophosphate, sodium tripolyphosphate and others; only here the disadvantage being their high cost.

Effervescent bath salts are categorized as fancy products rather than cosmetic. They are prepared with sodium carbonate and a crystalline acid. Citric and tartaric being the most commonly used.

METHOD OF MANUFACTURE

The carbonate and the acid are mixed and perfumed, coloured, wetted with alcohol and granulated by passing through a screen of a suitable mesh. The wet granules are then put in trays and the trays into a steam oven, so as to dry them with a small quantity of live steam. This results in a reaction on the surface with the formation of a neutral salt that acts as a protective coating against moisture in the air. Isopropyl alcohol of 99% strength is used for granulating because of its low water content and alcohol soluble gum binders but only in very small quantities.

The granules are ready for packing either as they are or in the form of one-bath tablets.

Phosphates can be included in the formula but one has to bear in mind the fact that large quantities of it would weaken the effervescence. Powdered detergents like sodium lauryl sulfate can be added but would put the preparation into a different category of bath preparations namely bubble bath preparations.

This is a formula of an effervescent bath preparation.

Formula 4

Sodium bicarbonate	45.0%
Tartaric acid	37.5
Sodium hexametaphosphate	10.0
Carboxymethyl cellulose	2.0
Sodium lauryl sulfate	5.0
Perfume	0,5

Procedure

Combine the dry ingredients in powdered form and mix until uniform. Dissolve the perfume in enough isopropyl alcohol to moisten the batch and add to the granulate and sieve through screen, and dry. The finished product is packed in airtight containers or pressed into tablets, preferably about 1¼ in. in diameter and ¼ in. thick, which are placed in the airtight containers.

BATH OILS

Bath oils too like bath salts are classified into two main types: one the oily type and the other either soluble or emulsifiable with water.

The first type is perhaps less popular. The product revolves around say, castor oil, alcohol for viscosity adjustments, perfume oil (about 3%) and an oil soluble colour. A small quantity of this oil is poured into the tub. The perfume oil is absorbed by the water and the other oil floats on the water surface as a thin film. This film is "absorbed" on the body while bathing.

Other oils can be used but present difficulty in colouring and perfuming. The one big advantage with castor oil being its easy miscibility in alcohol to give desired viscosity.

Synthetic oils like isopropyl myristate or plamitate, butyl stearate and others easily perfumed and coloured are suggested alternatives. Moreover, they are free from fatty odour and do not easily become rancid.

The percentage of alcohol content is directly related to the viscosity desired. These products do not possess detergent water softening or cleansing properties. The two formulas below typify such preparations.

Formula 5

Castor oil	30 to 80%
Ethyl alcohol	60 to 10
Perfume	10
Colour	q.s

Formula 6

Isopropyl myristate	90 to 95%
Perfume	5 to 10
Colour	q.s

Since isopropyl myristate or any other fatty acid easter may not absorb all perfume oil, alcohol or any other solvent may be used to obtain a clear solution.

Bath oil with detergent and water softening properties is a popular choice. It was previously made with 50 % sulfated oil (castor or soya) with suitable dosage of perfume and colour.

Bath oils made more recently are based on synthetic nonfoaming detergents. Sometimes water is used as diluents to reduce the cost. A perfume-solubilizing agent is used for a clear product.

METHOD OF MANUFACTURE

The manufacture of this type of bath oils is easy. The sulfated oil or the detergent is placed in a tank. The perfume and the solubilizing agent, if required, as a mixture is added to the oil and thoroughly stirred. Colour is added. The colour should be tested for compatibility or in other words it should not react with the product. The selection of the sulfated oil is done in such a manner wherein it accepts the addition of alcohol and water. Some sulfated castor oils and some sulfated soyaben oils take 10-20% perfume oil without a solubilizer and remain clear.

Here are some formulas based on sulfated castor oil

Formula

	7	8
Castor oil, sulfated	97%	50%
Ethyl alcohol	—	10 to 0
Solubilizer	—	7
Perfume	3	3
Colour	q.s	q.s
Water	—	30 to 40

A bath oil based on a sulfated detergent and non foaming type is shown in the formula below.

Formula 9

Sulfated detergent	30 to 40%
Water	60 to 50
Solubilizer	7
Perfume oil	3
Colour	q.s

The solubilizing agent selected here has to be carefully adjusted considering the perfume oil used as well as the solubilizing effect of the detergent.

BUBBLE BATH

Bubble bath are easily the most popular among the bath preparations.

They are available in different physical forms: powder, liquid, tablets, capsules, crystal and "soap cakes".

A good bubble bath should cater to the following requirements.

1. Should bubble easily without exersion of excessive water pressure.
2. Foam to be stable in hand and soft water as well as in the presence of soap.
3. It should prevent formation of a ring around the bathtub.
4. The recommended dilution in the tub should not irritate the skin or mucuous membranes.
5. It should contain ingredients that are low in cost to enable reasonable pricing of product in attractive packaging.

One would assume that only synthetic detergents are useful in the formulation of bubble baths, however the use of non detergent surface active agents, emulsifiers and sequestering agents is not uncommon.

Water-soluble gums, natural as well as synthetic, are also used for one very important property they possess; preventing dispersed soil form redepositing on the body by holding the soil particles in suspension. Not all of the good foaming detergents have this capacity. The gums if well formulated will also provided desired thickness to the liquid bubble bath.

Powdered bubble bath preparations contain non-hygroscopic detergents, foaming agents, and water softeners. Many of the alkyl aryl sulfates and sulfated fatty alcohols can be used for such a preparation. However, it must be seen that they do not cake in high humidity or become damp when used in combination with another material even at relatively low humidities although they may be excellent agents themselves.

Sodium chloride may be used as a diluent to keep the finished products free flowing especially when stored in large containers. Sodium polyphosphate or the amide of an alkyl phospate is used in place of sodium chloride when the product is more concentrated and used in smaller quantities.

Formulas 10 and 11 are some examples of bubble bath preparations with sodium chloride and 12 with substitute for sodium chloride as already discussed.

Formulas for Bubble Bath

	10	11	12
Sodium lauryl sulfate	30%	—	—
Sodium lauryl sulfoacetate	—	20%	60%
Sodium haxametaphosphate	5	—	—
Sodium carboxymethyl cellulose, low visc.	2	—	—
Sodium chloride, fine granular	60	47	—
Alkyl benzene sodium sulfonate	—	30	—
Sodium tripolyphosphate	—	—	38
Perfume	3	3	2

Procedure

Mix the dry ingredients until uniform, then add the perfume, and mix again until uniformly distributed. If diffculty is experienced with nonuniformity of odour, the perfume should first be dissolved in a little alcohol.

Generally the powdered bubble bath preparations are put in tablet form (rather easily at low humidity). A good bubble bath soap cake is obtained by way of compression of sodium lauryl sulfoacetate alone or in a mixture with boric acid, talc, starch and other materials.

But the liquid bubble bath is the most popular product. Earlier it was based on liquid soap such as potash and triethanolamine soaps of coconut oil, phosphate and other sequestering agents like the sodium phosphate, sodium hexametaphosphate, tetra sodium pyrophosphate, and sodium tripolyphosphate were added to keep insoluble soaps from anti foaming activity in. the bath. The tetrasodium salt of ethylene diamine tetra acetic acid is excellent in preventing the killing of foam by magnesium and calcium ions.

More modern bubble baths are based on synthetic surfactants which give good foam whether they are good detergents or not. Because by adding a good detergent both the qualities can be obtained. A majority of the good foamers also posses good detergent activity resulting in a good liquid bubble bath.

The ingredients of a bubble bath need to be selected in such a way that they remain as clear solutions over a wide range of temperature.

Formulas 13 to 18 provide excellent bubble bath preparations.

Formulas for Bubble Baths

	13	14	15	16	17	18
Fatty acid amine condensate	40%	30%	20%	—	—	—
Alkyl sulfonate	—	15	—	—	—	—
Sodium lauryl sulfate	—	—	20	30%	—	—
Triethanolamine lauryl sulfate	—	—	—	—	30%	—
Alkyl aryl polyethyl ether	—	—	—	—	—	07%
Diethyl ester of sodium succinic acid	—	—	—	—	—	06
Ethyl alcohol	—	—	—	—	—	10
Water	57	52	57	67	67	q.s 100
Perfume	03	03	03	03	03	03

BATH POWDERS

Bath powders of the past were based on corn and rice starch and boric acid. But both powders of the present constitute mainly any where between 60% - 90% of talc. Many other ingredients in smaller quantities are used ot give it the desired properties.

For instance these are:

Ingredients	Purpose
Chalk	for bulk and density
Zinc oxide	for bulk and density
Magnesium carbonate 5%	to adsorb perfume and hold it
Kaolin	to help powder to hold on to the body
Titanium dioxide 1-2%	to give some opacity to talc
Zinc stearate	to give desired slip to the product

Colour

Normally earth oxide colours are used. For example a mixture of cosmetic grade iron oxide (light pink colour) and light tanochre will produce a flesh coloured tint. This is added as an extender after all other ingredients are uniformly mixed. The extender should consist of 25% of dry colour and 75% talc uniformly mixed and then micropulverized.

Formulas for Bath Powders

	19	20	21	22
Talc	80%	70%	90%	60%
Magnesium carbonate	05	05	04	05
Kaolin	05	07	—	—
Titanium dioxide	02	—	—	02
Zinc stearate	07	04	05	10
Zinc oxide	—	03	—	12
Chalk, heavy	—	10	—	10
Perfume	01	01	01	01
Colour and extender	q.s	q.s	q.s	q.s

Procedure

Place the dry ingredients, except the magnesium carbonate in a powder mixer and mix until uniform. If a perfume fixative is to be added to counteract the earthy odour, it is incorporated at this point in the form of a 1% alcoholic solution. Mix the alcoholic solutions into the dry ingredients, until uniformly distributed. Mix the magnesium carbonate and the perfume in a separate container and sift into the batch while mixing. If colour is desired, add the extenders at this time and mix the batch until the colour is uniformly distributed. Then run the batch through a micropulverizer or through a bolting screen, and it is ready to be packaged. The bath powder should be packaged in a large diameter, round or square, flat box, so that a large puff can be enclosed.

15

Rouge

ROUGE IS SMALL but an important cosmetic from a lady's point of view. Its purpose is to simulate the rosy freshness of the young and healthy skin.

Table 1. Physical Forms of Modern Rouge

Form	Composition	Advantages	Disadvantages
Liquid			
Suspension	Pigment suspended in water alcohol, glycerol and other liquids	Cheap	Lacks cosmetic elegance; may require shaking before use
Emulsion	Pigments suspended in a fluid emulsion	Relatively cheap blends easily	Emulsion may separate; limited choice of suitable pigments
Cream			
Anhydrous cream	Pigmented ointment	Stable, easy application	May be greasy, may exude oil on standing and during changes of temperature
Water-containing	Pigmented cream	Pleasing appearance, easy blending	Limited shelf life through loss of water
Solid			
Waxes and oils	Similar to lipstick	Stable, easy to apply	Too much like lipstick

| Dry compact | Pigmented powder compressed with the help of a binder | Long-lasting, applies with puff, matte finish | Dusty, may crack and crumble |

LIQUID ROUGE

Suspension Type

The dye solution one can say was the precursor of rouge. But the tendency of dye solution to stain the skin and the difficulty in application lead the chemist to focus on pigmented solutions as simple means of applying a touch of red to the cheek. Such suspensions in their simplest forms settle rather quickly and need a thorough shaking before use. In order to retard this settling of the pigments and inert solids and to disperse, various suspending agents such as carboxymethyl cellulose, polyvinyl pyrrolidone and polyvinyl alcohol have been selected to act as colloids around the solid particles.

Another alternative to prevent the inert and white particles from settling is to have a voluminous precipitate floating in the water phase. This can be achieved by grinding zinc stearate into the water phase to yield a product, which gives the appearance of a lotion and provides a matte finish on application.

A widely followed way of doing this is by adding glyceryl or propylene glycol monostearates at high temperatures to the aqueous pigment suspension. The monostearate being dispersible in hot water, innumerable microscopic stearates start cooling down which in turn act as cushions to the heavier pigment particles.

Formulas 1-3 represent rouge solutions and formulas 4-6 describe rouge suspensions.

Formulas for Liquid Rouge

	1	2	3
Carmine NF	2.0%	—	—
Glycerol	5.0	—	—
Water	88.9	89.4%	—
D&C Red no.28	4	0.5	6%

Carbowax 400	—	10.0	6.0
Alcohol-soluble dye	—	—	4.0
Castor oil	—	—	10.0
Ethyl alcohol	—	—	74.0
Methyl p-hydroxy benzoate	0.1	0.1	—
Perfume	q.s	q.s	q.s

	4	5	6
Propylene glycol	3.0%	—	—
Polyvinyl alcohol 2%soln	2.5	—	--
Sorbitol hexaacetate,2% soln	2.5	—	—
Titanium dioxide	2.0	—	—
Chalk	4.8	—	—
Zinc stearate	2.4	17.0%	—
Colour	3.0	6.0	6.0%
Water	74.7	75.5	76.9
Zinc oxide	5.0	—	—
Glycerol	—	1.0	4.0
Dihydroacetic acid	—	0.5	—
Methyl p-hydroxy benzoate	0.1	—	0.1
Diethylene glycol monostearate	—	—	9.0
Spermaceti	—	—	3.0
Sodium lauryl sulfate	—	—	1.0
Perfume	q.s	q.s	q.s

Emulsion Type

The emulsion type rouge provides a blend of easy application and desirable appearance in the bottle. The only hitch being requirement of an acceptable shelf life for which careful formulation is necessary. The choice of colours and toners is restricted due to the alkaline nature of the emulsion. Formula 7 is one such example. The consistency of the lotion can be varied by varying the soap content or by adding quince seed extract, bentonite, carboxymethyl cellulose, or other thickening agents.

Formula 7

Mineral oil	39.4%
Oleic acid	7.3
Dry ingredients	
Titanium dioxide	0.6
Zinc stearate	0.42
Pigment	0.42
Aluminium hydroxide	0.36
Water	47.7
Triethanolamine	3.7
Methyl *p*-hydroxy benzoate	0.1
Perfume	q.s

Procedure

Heat the mineral oil and oleic acid to 60°C. Heat the water and triethanolamine to 60°C. Pregrind the dry ingredients and either roller-mill the latter into the oil phase, which is then converted into the oil emulsion or add the water phase to the oil phase, and add the preground dry ingredients to hot finished emulsion, stir while cooling, and perfume at 45°C.

The use of nonionics as emulsifying agents allows larger choice of colours but here careful balancing of ingredients is necessary to insure a long shelf life. Formula 8 is one such example.

Formula 8: Creamy Pearl Rough

Part 1

Lanolin oil	2.2%
Isopropyl myristate	2.7
Bees wax	2.7
Sorbitan stearate	5.5
Lanolin alcohol	5.3
Glyceryl tribehenate	3.9

Paraffin	4.4
Propyl paraben	0.1

Part 2

Water, distilled	44.8
Polysorbate 60	2.6
Propylene glycol	5.4
Methyl paraben	0.2
Imidazolidinyl urea	0.2

Part 3

Mica (and) Titanium dioxide (and) Carmine	6.7
Mica (and) Iron oxides (and) Titanium dioxide	6.6
Mica (and) Titanium dioxide	6.6

Part 4

Fragrance	0.1

Procedure

Heat Part 1 to 85°C. Heat Part 2 also to 85°C. Add premixed Part 3 to Part 2 with stirring. Add Part 1 with stirring and remove from heat. Add Part 4 with stirring at 45°C.

Cream Rouge

Cream rouge enjoys preference over other forms of rouge because of its ease of application, soft texture, perfect appearance, and satisfactory blending with powder.

Anhydrous Creams

There are two types of anhydrous rouge creams. One based on mineral oil and waxes which is the older version and the newer one which is based on lower alkyl fatty acid esters and carnauba wax to harden it. They present good shelf life with acceptable cosmetic properties. The first type is cheap but the key lies in the selection of raw materials so that it remains stable at temperatures above 40°C.

The base materials used not only render easy application but also insure heat stability. This type of rouge gives a greasy feeling unless sparingly used.

The second which is all oil-and-wax type is based upon the fact that carnauba wax, in the presence of talc, chalk, kaolin, and pigments stiffens isopropyl palmitate and other esters to the desired consistency. The esters being oily liquids with low viscosity yield very pleasing thin films on the skin. This mixture of carnauba wax and the esters, which is stiff and solid when direct pressure is applied, becomes fluid when rubbed gently with a circular motion. Moreover, a well-made mixture of this type of cream rouge can withstand temperatures above 50°C.

However, both types of creams tend to exude oil in tiny droplets when subjected to abrupt temperatures changes. This situation can be averted with the addition of beeswax, ozokerite, or lanolin or even a small amount of mineral oil. Infact any material acting as a third solvent to reduce brittleness also reduces sweating of the rouge.

The stability feel of the anhydrous cream rouges has a lot to do with pouring temperatures and cooling rates. Flaming improves the appearance of the surface, which is done after the cream has cooled down and set. If this is not observed the pigment settles slightly leaving a mottled surface deficient in pigment.

Cream rouges need to be perfumed with great care since they tend to develop an odd odour and since they are used sparingly the perfume must be retained for a long period.

Some examples of anhydrous cream rouge formulas are given below.

Formula 9

Petrolatum	76%
Mineral oil	8
Lanolin	4
Zinc oxide	5
Pigments	7
Perfume	q.s

Formula 10

Carnauba	6%
Ozokerite	10
Mineral oil	24
Isopropyl palmitate	27
Talc	10
Titanium dioxide	20
Colour	3
Perfume	q.s.

EMULSIFIED CREAMS

One major draw back of anhydrous cream rouge is its greasiness. To overcome this problem, cream rouges that are based on pigmented emulsions have been formulated. Almost all types of cream can be adapted as rouge bases. A well formulated emulsified rouge cream: (a) is easy for application, (b) has minimum drag (c) is non-greasy (d) is cooling, (e) has a pleasant texture and presents a subtle attraction to many women. A poorly formulated cream on the other hand may have mottled effect, may feel spongy because of occluded air, have a lard like appearance due to poor emulsification or bad choice of ingredients. Insufficient or unsuitable humectants may cause the cream to dry up and loose its customer appeal. Some examples of emulsified cream rouge:

Formula 11 (Vanishing Cream Type)

Stearic acid	20.0%
Cetyl alcohol	2.0
Glycerol	10.0
Potassium hydroxide	1.0
Water	58.9
Pigment	8.0
Methyl p-hydroxy benzoate	0.1
Perfume	q.s.

Procedure

Heat the stearic acid and cetyl alcohol to 70°C. Heat the glycerol, potassium hydroxide, and water to 70°C. Add the aqueous phase to the oil phase. Continue to stir, while cooling add perfume at 45°C. Then remove a small amount of the cream, add the pigment, and mix it in, or use a loose-ointment mill and stir it back into the bulk of the cream.

Formula 12 (Cold Cream Type)

White beeswax	12.0%
Petrolatum	24.0
Spermaceti	8.0
Mineral oil	22.0
Borax	0.8
Water	30.0
Pigment	3.1
Methyl p-hydroxy benzoate	0.1
Perfume	q.s.

Procedure

Follow same directions as for Formula 11, except that the oil phase consists of beeswax, petrolatum, spermaceti, and mineral oil, water phase consists of borax and water.

Formula 13 (Neutral Cream Type)

Arlacel 83	4.0%
Lanolin	4.0
Mineral oil	14.0
Petrolatum	28.0
Sorbitol syrup	4.9
Water	35.0
Pigment	10.0
Methyl p-hydroxy benzoate	0.1
Perfume	q.s.

Procedure

Follow the same directions as for Formula 11, except that oil phase consists of Arlacel 83, lanolin, mineral oil, and petrolatum; water phase consists of sorbitol syrup and water.

Formula 14

Glyceryl monostearate (self-emulsifying)	11.9%
Spermaceti	4.5
Glycerol	4.5
Water	71.0
Colour	8.0
Methyl p-hydroxy benzoate	0.1
Perfume	q.s.

SOLID ROUGES

Solid Oil and Wax Rouges

The idea of having rouges in the form and composition similar to lipsticks has been cropping up from time to time. Actually some women do use their lipsticks as substitutes for rouges. Commercially it has not been offered because of the confusion it may create in the customer's mind.

As a matter of fact any lipstick of preferably creamy texture and free of halogenated fluorescence which does not permit delicate blending of rouge especially on an already creamed face would be acceptable. A careless application of this form of rouge may however result in the "apple cheek" variety.

A simple example of an oil and wax rouge base is given below.

Formula 15

Castor oil	77.4%
Candelilla wax	9.9
Carnauba wax	2.7
Colour	10.0
Perfume	q.s.

Another tested solid oil and wax rouge formulation using isopropyl esters and carnauba wax and sufficient colour is given below.

Formula 16

Isopropyl myristate	45.0%
Mineral oil	22.5
Carnauba wax	8.8
Lanolin	4.0
Colour	19.7
Perfume	q.s.

COMPACT ROUGES

Compact rouge is the most popular form of rouge and is essentially a pigmented powder compressed to a firm cake with the help of a binder. However, it not only requires a sound technical know how but also specialized equipment.

The earlier version of rouge cake was made by pouring wet powder containing gypsum into moulds where the rouge set like plaster of Paris. They were hard and gritty compared to the modern cakes. These cakes are soft and smooth and can be produced cheaply in large quantities.

A good dry compact rouge exhibits the following characteristics.

1. Smooth texture.
2. Perfect distribution of colour.
3. Good covering power.
4. Very small particle size.
5. Perfect blending.
6. Ease of application.
7. Good adhesion.
8. Ease of removal without trace or residue.

In addition the cake should not flake, crumble, crack, or be too hard.

Some common defects observed in compact rouges:

1. Being powdery, dusty and irritating.
2. Hard and not readily blended.
3. To glaze when rubbed with a puff.

The raw materials used in regular base face powder are same, but differ in ratio as far as compounded compact rouge is concerned.

A study of the commonly used raw materials for the manufacture for compressed rouge powder is done here below.

Talc

Should possess slip without greasiness; it should be free of "Sparklers", have uniform small particle size. It is the main ingredient for an easy application. If the content in rouge is too high it gives a thin glassy appearance. The talc here may not be pure white since the product is strongly pigmented.

Kaolin

Kaolin is soft and opaque on application. The colloidally refined form of kaolin can be used for bulking and to adjust the overall fluffiness of the composition. It has bonding properties because of its capacity to absorb and retain certain amount of moisture. However, when used in a high percentage, it gives spots and streaks due to its hygroscopicity. Kaolin causes many perfumes to deteriorate and for this reason it has to be used cautiously.

Chalk

This material has no major role in the formulation of compact rouge except in small quantities as a perfume carrier. It has a tendency to make the compact brittle.

Magnesium Carbonate

It is added as a bulking agent and as a carrier of perfume. It has very little covering power and too much of it makes the compact fragile.

Zinc Oxide

Its usual concentration varies from 5-30% depending on the concentration of other materials. Its white pigment gives opacity without blue under tones. Brightens red tones and enhances adhesion of rouge to skin. But it reduces the strength of the compact.

Titanium Dioxide

It has 4-5 times more hiding power than zinc oxide. It gives lively shades however with blue undertone.

Metal Stearates

Zinc, magnesium and aluminum stearates are almost essential for a well-compounded rouge and are used in concentrations between 3-10%. They are useful in providing adhesion to skin, smoothness to the cake and act as binders in conjunction with gum binders. However, colour streaking may occur when used in high concentrations.

Starch

Dry starch is useful in small quantities. When used in conjunction with aqueous binders it may cause swelling, cracking and breaking of rouge cake. It can however be used in the binding solution which needs to be preserved.

Pigments

Water and oil soluble dyes are usually avoided. Colour lakes, together with organic toners, reduced toners and to a lesser extent inorganic oxides, are the principle pigments used and they also provide a wide range in shade and brilliance.

Perfume

A discreet odour is recommended for compact rouge bearing in mind dry rouge, which is sparingly used, and it may last for years. Because of its dense compact form perfume is released very slowly.

Preservatives

A gum binder being used, it is advisable to incorporate a preservative.

Binder

Actually a successful compact rouge hinges heavily on the selection of the binder. Several binder combinations have been proposed and used. Some of used ones are given hereunder.

Water soluble binder.
Water repellent binder.
Emulsion binder.
Dry binder.

Each one has advantages and disadvantages and is necessary to make an individual study to be able to put them to best use.

Water Soluble Binder

This type of binder is basically a gum solution and both natural and synthetic gums are employed. Gum tragacanth, gum arabic, gum karaya, quince seed and Irish moss extracts are all natural ingredients. A combination of gum tragacanth and quince seed has been used. The use of gum arabic needs caution because it may give the cake a very hard consistency. Being natural products, it is difficult to obtain material that is pure, uniform and low in bacteria and mould count, although standardized grades are commercially available.

As a result of these problems, manufacturers resorted to synthetic gums like methyl cellulose, carboxy methyl cellulose, polyvinyl pyrrolidone and others. The concentrations of gum used varies from 0.1% to 3%. Starch solution is used by some as a binder as well.

Water Repellent Binder

Because compact rouges are subject to spotting by water, water repellent binder came into the picture. These binders can be liquids, semi-solids, or solids, which are incorporated into the rouge in the molten state. The following materials either alone or in combination can be used: Mineral oil, petrolatum, fatty acid esters, lanolin, and lanolin derivatives in concentrations varying from 0.2%-2%. This quantity of fatty material is insufficient to form a solid cake and requires about 10% of water, which is mixed with the powder before pressing. A small quantity of wetting agent ensures a uniform water-powder mixture.

Emulsion Binder

The difficulty in uniform distribution of small quantities of oil or waxes into preground rouge powder led to the choice of an emulsion binder. A thin emulsion has bulk as well as water phase which make pressing easy. Moreover, an emulsion incorporated into a powder prevents moisture loss and makes the manufacturing procedure smoother. Uniform distribution of oil phase in the powder prevents glazing and lumps which are likely to occur when the oils alone are added to the powder.

The use of soaps as emulsifying agents could cause skin irritation although the adhesion of the powder is good. Sorbitol derivatives however, give acceptable emulsion and nonirritating binders.

Dry Binder

Dry binders like metallic stearates, in addition to pasty and liquid binders, provide alternative choice as binding agents for the rouge powder. However, the compression has to be increased for obtaining rouge compacts. Such a compact exhibits smooth texture and good adhesion to the skin but has shown a tendency to irritate skin sensitive to the alkalinity of some metal stearates.

THE MANUFACTURE OF DRY ROUGE

The basic steps involved in the manufacture of dry rouge are grinding, matching of shades, addition of binder, moulding and pressing. There are several ways of performing these operations and the choice of method depends upon the prevailing conditions and the particular requirements.

Grinding

The main purpose of various manufacturing procedures is distribution of colours. The older process called the French or Continental process is slow and elaborate. It yields excellent rouges with a minimum of equipment. In this method , the lakes, toners and the other pigments are first premixed in spiral ribbon or pony mixers. Then it is converted into a dough like paste by adding water or the binder solution. It is then placed in shallow trays either with or without previous granulation, for drying in ovens. The resultant hard, dry cakes are reground in hammer or ball mills. If this cycle of wetting, drying and grinding is repeated, a soft and brilliant rouge is obtained.

In the American process, colours are distributed by powerful grinding equipment which includes hammer mills with air classifiers, cyclone mills or attrition mills (refer to face powders chapter). Another efficient method is provided by edge runner mills which function on a principle similar to that of a pestle and a mortar. These mills mix and grind simultaneously. However, it is necessary to lighten the wheels to avoid squeezing the soft stearates into lumps which are difficult to disperse.

It is advisable to keep a check on the manufacturing procedure from time to time because grinding equipment wears and becomes less efficient; mills of the same type but of different capacities need not necessarily show same efficiency and their characteristics have to be known. The addition of small quantities of oil or water in the powder while grinding helps in better distribution of colours. Titanium dioxide being neutral gives more stable colour shades than the alkaline zinc oxide.

Shade Matching

For close matching, small amounts of dry powder have to be moistened with water or binder, ground through a laboratory mill, moulded, pressed and dried. This sample rouge is then compared with the master shade. Necessary adjustments have to be made on the dry powder. Further, it is advisable to prepare several concentration of blends to have freedom of adjustments.

Addition of Binder

It is important that the binder is evenly and completely distributed. The binder is sprayed into the powder while it is being mixed. Rouge powder being water repellent, and since most binders contain water, it is customary to add a wetting agent to ensure uniform distribution. Emulsion binder is preferred as it also acts as a wetting agent and helps in the distribution of the finely divided oil phase.

Moulding

Moulding and pressing operations are involved in the manufacture of tabletting type rouge. Normally before the compact rouge is pressed into shallow pans, it has to be moulded. It is done either manually or automatically either in single or multiple cavity moulds. The most important aim of moulding is to fill each cavity with the same amount of powder. However, it gives the compressed cake only approximate uniformity since the weight depends on the content of air in the powder. Regulated devices are necessary for moulding otherwise rouge compacts vary greatly in hardness and ease in rub-off. The powder is (like a dome) usually moulded in such a manner that a pronounced hump is created in the center, which is dense and hard. If it is done-with a level surface, the centre of the pressed cake is soft.

Pressing

What the pressing operations should do theoretically is

1. Expel the air from the interstices between the powder particles.
2. Bring them into physical content with each other.
3. Weld the binder film between particles.

The die should leave a small and definite clearance between itself and edge of the pan to allow passage to the expelled air.

Ideal pressing employs a fixed pressure cycle i.e., first compress gradually and continue till the optimum is reached; and disengage slowly and smoothly after sustaining the pressure briefly at the optimum. Foot-operated presses, air cylinders, and hydraulic cylinders are all suitable.

Tabeletting machines mould and press automatically but are rather expensive for the manufacture of rouge.

The thin rubber dam can be slipped over the die in order to prevent "lifting" of the compressed rouge. It does not work if rouge is embossed. The die in this case has to be finely polished to avoid "lifting".

It is preferable that the metal pans rest on resilient rubber cushion while being pressed because they are never perfectly flat and more over they should permanently be coated with tackly glue so that there is good adhesion between rouge cake and the metal pan even if the latter, somewhat bent during pressing springs back when pressure is relieved.

The spring room in which the rouge is being moulded and pressed should be air-conditioned for uniform temperature and relative humidity. Further, it is advisable to maintain the correct moisture content of the powder during moulding and pressing.

After the formation of the moulds the rate of drying should be uniformly slow to avoid a dry crust and an undesirable top cast.

An example of dry rouge powder is given below:

Formula 17

	%
Talc	48
Kaolin	16
Chalk	4
Magnesium carbonate	4
Zinc stearate	4
Titanium dioxide	12
Colour	12
Perfume	q.s.

Procedure

For directions, the reader is referred to preceding parts of the chapter, under the heading "The manufacturing of dry rouge".

16

Depilatories

THE DESIRE FOR removal of undesired hair for improving personal appearance, hygiene has haunted men and women since ancient times. It is recorded that the primitives used mineral, vegetable or animal matter either in ointment or in paste form for the removal of unwanted hair or to prevent its growth.

To quote a few of the ancient products: 1, burnt chaetopod boiled with balanites oil, 2. Burnt leaf of lotus in oil, 3. shell of the tortoise with the fat of the hippopotamus, the blood of oxen, asses, pigs, hounds and goats, together with stibium and malachite.

The original classical depilatory contained natural arsenic trisulfide, quick lime and starch, and was made into a paste with water. The term depilation is defined as the removal of human hair fiber, through chemical degradation with inorganic sulfides and organic thiols.

Recent fashions in women's apparel created a large demand for depilators. Shaving is usually, a bi-weekly procedure, and the process is made more convenient, by specially designed razors and elegant shaving creams. However, this procedure is not very well acceptable by women due to its disadvantages.

Next is electrolysis, which is not only costly but time consuming. More recently epilatory waxes or other similar adhesive compositions were introduced as depilatory creams. This approach is much more convenient and enjoys consumer acceptance.

EPILATORS

Epilants are hair pulling type in semisolid condition which are adhesive to skin.

(A) Wax-Rosin compositions are designed for application in the molten state to the hirsute area and allowed to solidify, so that the hair becomes enmeshed in the plastic mass. After keeping for 15-30 minutes removal of the waxy film from the site uproots and removes the hair.

(B) Adhesive semisolid compositions which are permanently stickly at room temperature and applied generally on a flexible supporting material such as fabric. Removal of the adhesive compositions by stripping removes the hair easily.

General Formula of Epilator

Rosin	69%
Bees wax	20
Burgundy pitch	4
Gum camphor	3
Oil of Bergamot	2
Oil of eucalyptus	1
Oil of skunk	1

The molten mass was poured into moulds to form sticks, which were heated before being applied to the hairy area, on cooling the solidified stick was removed quickly from the skin, thereby removing the embedded hair.

Later on this was improved and an attempt was made to even restrict further growth of hair through improvised preparations.

After epilating by Rosin-wax mixture another mixture of limewater, hydrogen peroxide, and oil of turpentine, with colour and perfume, was applied to the area in an attempt to restrict further hair growth. Formulations were changed by including viscous plasticizers like honey and non drying oils, such as mineral oil or olive oil to make the composition utilizable without preliminary heating.

Replacement with glucose, molasses, or honey and with water insoluble fillers in a ratio of 2:1 respectively, was proposed in addition to raw rubber in a volatile solvent.

The wax-rosin compositions were also supplied with a backing material which formed a flexible mounting for the adhesive mixture. Similarly adhesive compositions, containing zinc oxide, and various ketones were patented in Western Countries. Encompassing all the previous art, ready for application products were introduced. For example a ready to use and instantly removable depilatory pad type for application to the body, was formulated. It comprised of a combination of a substantially impermeable base section and a permanently coated stick mounted on the said base section. Any hair, coming in contact with such a pad comes off easily from the skin when removed without causing injury.

The advantage of such a system is that it can be handled/operated by even unskilled or trained persons. However, though it claims no allergic reaction or discomfort it is not 100% true.

Later several brands of these mechanical hair-removing aids in small cakes, canisters, kits, or tubes were available, but their popularity has been largely superseded by the newer chemical depilators.

Rosin Based Epilatories

Formula 1

Rosin	42%
Bees wax	3%
Carnauba wax	6%
Mineral oil	15%
Base	q.s.

Formula 2

Rosin	50%
Bees wax	24%
Petrolatum	4
Benzocaine	2
Base	q.s.

Formula 3

Rosin	64%
Bees wax	8
Carnauba wax	24
Lin seed oil	4

Formula 4

Rosin	58%
Bees wax	22
Ceresin	10
Base	q.s.

CHEMICAL DEPILATORIES

The removal of superfluous, undesirable hair was revolutionized by developing chemical agents either in paste or in cream formulary and they became popular in the evolution of variety of formulations.

An ideal depilatory formulation would possess the following qualities.

1. Should transform human hair into a soft plastic mass, easily removed by wiping or rinsing.
2. Should be non-toxic systemically and non-irritating to the skin even in the long run.
3. Should be easily applicable, economical to use, and stable in the tube or jar.
4. Should be cosmetically elegant, odourless, or pleasantly perfumed, white or natural in colour, non injurious to the skin, stainless on cloth.

However, hypothetically it sounds satisfactory but practically it is not foolproof. The reason being, the proteinaceous nature of skin and hair which is akin, and the treatment of hair with chemicals will equally affect the skin, which is undesirable.

Mechanism of Action

The chemical treatment of hair involves the rupture of hair, breakdown and reformation of sulfide linkage, which are responsible for stability and flexibility of the hair fibre. But at the same time, it should not be a complete breaking of all cross linking disulfide bonds, in the permanent waving process. The various chemical agents as alkali, metal sulfides, sulfites, cyanides, amines, mercaptans, and certain metal salts. The s-s bond is affected with the increasing osmotic pressure within the hair fibre and as a result of which it swells, loses its tensile strength and generally deteriorates. A mass of jelly like consistency, which can be easily removed by wiping or scraping, is the final stage of alkaline hydrolysis in the presence of a reducing agent. Both the outer layer (cuticle) and the inner colour bearing layer (cortex) are disintegrated.

FORMULATION WITH METALLIC SULFIDES

The early American patent literature on depilatories pertained to tablet, powder, soap, or paste compositions containing barium, strontium, and sodium sulfides or polysulfides.

In 1912 Stone disclosed a depilatory formulation containing calcium hydroxide, sodium sulfide and calcium sulfide hydrate. In 1921 Donner altered the formulations, by reducing the odour, increasing the stability and improving the cosmetic elegance, over the previous formulations be it a liquid, powder or lotion formula containing sulfides. Subsequently a stable transparent jelly composition, which was designed to give better visual control over the depilating process was patented in 1936. A 15% lithium hydroxide solution was saturated with hydrogen sulfide in the presence of an inert gas until the solution contained 8.25% of it by weight. This solution was added to a jelly base containing about 6% of tragacanth, karaya or locust bean gum. The resultant clear jelly was applied and then worn for half hour till complete depilation was indicated.

Another patent with methyl cellulose gel base reported good stability and rapid effectiveness containing sodium sulfide, calcium chloride and calcium carbonate. A Japanese patent described the

combination of starch paste, and hydrophilic ointment in the preparation of a strontium sulfide cream depilatory.

Formula 5: Powder Depilatory

Barium sulfide	31.0%
Titanium dioxide	18.0
Corn starch	50.5
Menthol	0.25
Perfume	0.25

Formula 6

Strontium sulfide	35.0%
Corn starch	35.0
Powdered soap	5.0
Zinc oxide	23.0
Benzocaine	0.2
Perfume	1.8

*Mix with water at the time of application.

Formula 7: Depilatory Pastes

Part A

Sodium sulfide	4%
Glycerol	1
Kaolin	32%
Water	63

Part B

Barium sulfide	8%
Calcium carbonate	32%
Powdered soap	4
Glycerol	2
Water	54

Part C

Strontium sulfide	30%
Zinc Oxide	8%
Glycerol	8%
Methyl cellulose	2.5
Water	51.5

Part D

Strontium sulfide	35%
Titanium dioxide	4
Glycerol	5
Menthol	2.5
Perfume	1.0
Water	52.5

Later on in France many patents were obtained on inventions based on the aliphatic mercapto acids and salts thereof. (Thioglycolic, Thiolactic and Thiocyanic). These patented compounds were powders, creams, jellies or liquids and permitted the use of a wide range of perfumery materials. Depilatories containing thioglycollic acid are commonly employed in media having a pH range of 10-12.5. Later modifications were made to reduce the pH to 8-10, in a suitable base containing anionic wetting agents and also the addition of pro-oxidants like manganese, iron, and copper salts helped in increasing the efficiency and safety of thioglycolate depilatories.

Subsequently soluble xanthates eg. sodium ethyl or methyl xanthates had been proposed along with sodium dodecyl xanthate as ·emulsifying agent and calcium thioglycolate booster in a cream base, as an emulsifying agent.

Practical depilation could be achieved on the formulation of the base, in which the alkaline thioglycolate is incorporated.

1. The paste, cream or lotion must be of proper consistency, capable of being localized at the site of application, easily spread and non drying.
2. It should hold moisture for 15 minutes after applications.

3. It should maintain "build up" around the hair shaft and cling to the hairy area. (This could be achieved by adding suitable surface tension reducing agents compatible with alkali and alkali earth metals either anionics like alkali metals fatty alcohol sulfates, alkyl aryl sulfonates and several non-ionics of the polyoxy alkylene alcohol or either types, as suitable wetting agents and provide satisfactory emulsifiers, for formulations of lotions and creams which contain high concentrations of electrolytes. Thickening agents like tragacanth, karaya, guar, and quince seed exracts polyvinyl alcohol, methyl or hydroxy ethyl cellulose can also help in maintaining consistency and emolliency is achieved with cetyl or stearyl alcohol.

Formula 8: Depilatories Containing Thioglycolates

Calcium thioglycolate trihydrate	6.0%
Calcium carbonate light USP	21.0
Calcium hydroxide USP	1.5
Cetyl alcohol flakes NF	4.5
Sodium lauryl sulfate USP	0.5
Sodium silicate solution	3.5
Perfume	0.5
Distilled water	q.s.

Procedure

Mix 4.5 gms of sodium lauryl sulfate in 15 ml of hot water (65°C) and make a solution. Add molten cetyl alcohol to this mixture while hot and agitate while cooling to form and emulsion. Add previously prepared emulsion to this slurry and agitate for 30 minutes at 40°C. In another vessel make a suspension by mixing the calcium hydroxide and calcium thioglycolate in 10 ml of distilled water containing 0.5 gm of sodium lauryl sulphate. Add this suspension to the previously prepared mixture of emulsion and calcium carbonate. Then agitate at 40°C. Add perfume and continue agitation for 30 minutes.

Add water if necessary to make up the necessary weight. Later on this product can be roller milled, to make it free from gritty crystals and entrapped air. Then the product is ready to be filled in wax lined tubes just before congealing temperature.

Formula 9

Mercapto acetic acid	2.4%
Strontium hydrate	10.0
Calcium oxide	2.4
Colloidal clay	20.4
Methyl cellulose	2.2
Perfume	0.16
Water	62.44

Formula 10

Calcium thioglycolate	15%
Calcium hydroxide	5
Calcium carbonate	60
Sodium lauryl sulphate	0.5
	80.5

Formula 11

Thioglycolic acid	8%
Calcium oxide	8
Sodium dodecyl xanthate	15
Purified sperm oil	5
Water	64

The above formulations have a problem of perfuming. The task of masking the unpleasant odour and rendering it consumer acceptable makes it an arduous and frustrating task. There is a possibility of the perfuming agents reacting with the ingredients, which may develop discoloration as well distortion of the original perfume. Hence they are to be selected carefully after intensive trials.

Packing of Thioglycolate Depilatories

The packing of depilatories can be done in jars, and tubes with wax lined tin or preferably lead for the bodies of collapsible tubes.

Either polyethylene or polyvinyl chloride containers are not fully satisfactory due to "cave in" of the side walls of round and oval polyethylene bottles. However, advance in plastic technology gives hope for suitable containers.

Toxicity

Although several cases of apparent sensitization of thioglycolic acid have been cited, toxicity experiments show that solutions containing less than 80% of purified thioglycolate apparently do not cause primary skin irritation.

ANALYSIS OF THIOGLYCOLIC ACID IN DEPILATORIES

Iodometric Method

The volumetric determination of thioglycolic acid is based on oxidation of the acid to dithioglycolic acid.

$$\begin{array}{ll} HS.CH_2COOH & S.CH_2COOH \\ & + I_2 = \quad | \qquad\qquad + 2\,HI \\ HS.CH_2COOH & S.CH_2COOH \end{array}$$

$$I = HS.CH_2COOH = 92.11$$

The determination can be done by titrating directly with 0.1 N Iodine, or the titration can be made with 0.1 N KIO3 after addition of KI to the acidified solution:

$$KIO_3 + 5\,KI + 6\,HCl = 6\,KCl + 3\,H_2O + 3\,I_2$$

$$I = \frac{KIO_3}{6} = 35.6'$$

The iodine that is liberated reacts as mentioned above.

Reagents

Hydrochloric acid, conc.
Potassium iodide (powder)
Potassium iodate 0.1 N KIO_3. Fisher reagent
Congo red paper
Starch solution or thyodene powder indicator

Procedure

Weigh approximately 10 g of depilatory on a watch glass. Transfer to a 500 ml beaker containing 100 ml water. Add 20 ml hydrochloric acid to liberate the thioglycolic acid. Solution should be strongly acid to Congo red paper. Warm the contents to 60-70°C for 3 minutes and cool to 50°C. Add approximately 1 g of potassium iodide KI. Titrate with 0.1 N KIO_3. Before the solution turns slightly amber (due to free iodine), add indicator and complete the titration to a faint lilac colour endpoint. The solution is not blue, due to the organic matter that is present.

$$\%\text{Thioglycollic acid} = \frac{9.211(A)(N)}{W}$$

A = ml potassium iodate used in the titration
N = Normality of potassium iodate
W = Weight of sample in gram

17

Shelf Life (Stability) of Cosmetics

LIKE ANY OTHER pharmaceutical product, cosmetics too need shelf-life guarantee, without any deterioration and in order to maintain this preservatives are suggested, which will help preventing microbial growth or retarding oxidation in fats and oils etc. However, a preservative is different from germicides or antiseptics. The problems in shelf-life cosmetics are diverse. For example:

1. May be due to the physical structure of the preparation like, liquid emulsions, ointments, powders, etc., may be due to factors influencing the growth of microorganisms like nutritional factors, moisture, pH, temperature etc.,
2. Factors affecting the action of the preservative concentration, incompatible substances, nature of combination etc., of incompatibilities.
3. Presence of compounds, which are inherently bacteriostatic or fungistatic.
4. Purity of raw materials and maintenance of cleanliness/sanitary operations while manufacturing, packaging of the product.

Thus it is evident that deterioration may be due to physical, chemical, microbial, enzymatic conditions. Sometimes the deterioration is visible physically, or organoleptically like odour, colour, texture etc. Odour changes may be due to the production of volatile substances, such as aldehydes, keytones, acids, amines, sulfides, mercaptans etc. Similarly colour changes are due to growth of pigment producing organisms to oxidative reactions, and other chemical reactions.

Changes in texture may be due to hydrolysis of starches, solubilisation of proteinaceous material and microbial over growth. Additionally sometimes destruction of active ingredients such as vitamins, hormones, and allied substances may happen.

Most of the cosmetics, especially containing water are susceptible to microbial growth. All most all cosmetics containing water are susceptible to deterioration/microbial action. Sometimes, even dry products like lipsticks too. Thus it is clear that all types of cosmetics preparations, need preservatives to maintain their shelflife unless otherwise proved, that they are resistant to the growth of microorganisms.

Often selection of preservatives poses problems. An ideal preservative must have the following qualities.

1. Should be effective in all conditions of environment and with a variety of ingredients with incompatibility and must not alter the pH of the preparation.
2. It is better if it is soluble in its effective concentration
3. It must be stable and capable of sustained action.
4. Must be colourless and odourless.
5. Easily and economically formulated in the product.
6. Should be non-toxic and must not produce irritation.

Specially for cosmetics, this quality is highly essential. Recent preservatives like parabens etc have dominated (superceded) the old preservatives like salicylic acid, boric acid and its derivatives.

In general, the organisms that effect the cosmetics are Penicillium, Aspergillus, Rhizopus, and Mucoir and Botrytis cinerca. Vegetative growth is fluffy, grey white to dark, with small single celled, spores, borne on crowded, fine, warty swellings, along the rounded tips of upright branches. This is visible on creams where it subsists on the fatty acid residues of soap substances. The other possible organisms are Arternaria, Stemphylium, Cladosporium, Yeasts,

Bacteria etc. and in bacteria, B-mycoides, Aerobacter acrogens, Pseudomonas Sp. etc.

FACTORS INFLUENCING THE GROWTH OF MICROORGANISMS

The general substrates for microorganisms found in cosmetics can be classified as:

1. Carbohydrates and glycosides: Natural gums, mucilages, pectins, starches, dextrins and sugars.
2. Alcohols: Glycerol, mannitol and fatty alcohols.
3. Fatty acids and their esters: animal and vegetables fats, oils and waxes.
4. Steroids: Cholesterol, ergosterol and lanolin.
5. Proteins, peptones and aminoacids.
6. Vitamins.

Minerals

As minerals are essential to microbial growth although they are highly variable in their requirements for growth and at times antagonize biological effects. A single substrate like gelatin, may provide all of the above requirements.

Moisture Content

Bacteria in general require higher water content. Other factors essential for growth of microbes are pH, temperature, oxygen etc., surface active agents generally present in cosmetics, may have an effect on the type of growth of certain organisms. Among the most obvious effects is the tendency to grow large in size. The effect of ingredients such as certain perfumes, and essential oils, may provide added protection.

Efficacy of Preservatives

A preservative which can act with minimum concentration and

maximum efficacy is ideal or the effective concentration should be less. The solubility of preservative will also help in improving their efficacy. For e.g. hydroxybenzoates. pH also has a considerable effect on preservatives.

Evaluation of Preservatives

The official A.O.A.C phenol coefficient methods, utilizes salmonella typhosa, and micrococcus pyogenes var. aureus as test organisms. Another method that can be followed is "zone of inhibition".

Procedure/Method for Testing

1. Keep the finished cosmetic in the final container at room temperature for 6 months and examine periodically.
2. Inoculate the bacteria with various typical bacteria, yeasts, moulds, and incubate for several weeks, examining the product macroscopically and microscopically for evidence of growth.

List of Preservatives

1. Organic acids
2. Alcohols
3. Aldehydes
4. Essential oils
5. Phenolic compounds
6. Esters of hydroxy benzoates
7. O phenyl phenol
8. Mercury compounds
9. Surfactants
10. Miscellaneous nitrogen compounds

Antioxidants

Cosmetic preparations containing fats and oils, particularly those characterized by a high-percentage of unsaturated linkages, are susceptible to oxidative degradation. These materials may develop rancidity. The addition of antioxidants prevents rancidity, as they retard oxidation and minimize changes in colour and texture of the product. Moreover they can retard the breakdown of certain active constituents, such as vitamins. Common antioxidants that are generally recommended are BHA, nordihydroguaiaretic acid and phosphoric acid. A commercially available combination of butylated hydroxyanisole (20%) and 6% of propyl gallate, 4% of citric acid and 70% of propylene glycol, is branded as Tenox II. Sustane is a popularly used preservative in the food industry, but can also be used in cosmetics. Other common anti-oxidants are butyl hydroxy quinone, propyl gallate which protects against the pro-oxidant effects of dissolved iron.

As there is a strong possibility for the micro-organisms to grow in fatty media, with subsequent splitting of fatty glycerides into glycerol and fatty acids, followed by further decomposition. Moulds, principally the pencillin aspergilli, attack fats and thereon rancidity develops. However, adequate amounts and types of preservatives prevent this type of rancidity.

LIST OF ANTIOXIDANTS

Phenolic Type

Amyl gallate	Dihydroxyphenols
Butylated hydroxyanisole (BHA)	Gallic acid
Butylated hydroxytoluene (BHT)	Guaiacol
2,5-di-tert-Butyl hydroquinone	Gum guaniac
Nordihydroguaiaretic acid (NDGA)	Propyl gallate

Quinone Type

Hydroxycoumarins
Tocopherols
Solvent-extracted wheat germ oil

Amine Type

Casein and edestin

Ethanolamine

Glutamic acid

Hydroxamic acids

Kephalin

Lecithin

Plant and animal phosphatides

Purines (Xanthine and uric acid)

Organic Acids, Alcohols and Esters

Ascorbic acid

Citric acid

Dilauryl thiopropionate

Distearyl thiopropionate

Glucuronic acid

Isopropyl citrate malic acid

Malonic acid

Mannitol

Oxalic acid

Propionic acid

Sorbitol

Tartaric acid

Thiopropionic acid

Inorganic Acids and their Salts

Phosphoric acid and its salts fall in this category.

Index